CAMBRIDGE LIBRARY COLLECTION

Books of enduring scholarly value

Technology

The focus of this series is engineering, broadly construed. It covers technological innovation from a range of periods and cultures, but centres on the technological achievements of the industrial era in the West, particularly in the nineteenth century, as understood by their contemporaries. Infrastructure is one major focus, covering the building of railways and canals, bridges and tunnels, land drainage, the laying of submarine cables, and the construction of docks and lighthouses. Other key topics include developments in industrial and manufacturing fields such as mining technology, the production of iron and steel, the use of steam power, and chemical processes such as photography and textile dyes.

Economics of Construction in Relation to Framed Structures

From the 1850s onwards, the civil engineer Robert Henry Bow (1827–1909) became known for his expertise in structural analysis, publishing on the design of bridge and roof trusses, and working with the prolific railway engineer Sir Thomas Bouch (of later Tay Bridge infamy). In the first part of this 1873 publication, Bow describes 337 different truss structures, grouping them into four classes according to their structural characteristics: statically determinate, kinematically determinate, indeterminate, and other. In the second part, he describes a method for graphically analysing truss structures, based on the work of Thomas Maxwell and others, and applies this method to the structures listed in the first part. Perhaps of most interest to the working engineer are the explanations as to which structures are most efficient given typical material constraints, such as girders of uniform cross-section. The work remained a useful resource for practising engineers well into the twentieth century.

Cambridge University Press has long been a pioneer in the reissuing of out-of-print titles from its own backlist, producing digital reprints of books that are still sought after by scholars and students but could not be reprinted economically using traditional technology. The Cambridge Library Collection extends this activity to a wider range of books which are still of importance to researchers and professionals, either for the source material they contain, or as landmarks in the history of their academic discipline.

Drawing from the world-renowned collections in the Cambridge University Library and other partner libraries, and guided by the advice of experts in each subject area, Cambridge University Press is using state-of-the-art scanning machines in its own Printing House to capture the content of each book selected for inclusion. The files are processed to give a consistently clear, crisp image, and the books finished to the high quality standard for which the Press is recognised around the world. The latest print-on-demand technology ensures that the books will remain available indefinitely, and that orders for single or multiple copies can quickly be supplied.

The Cambridge Library Collection brings back to life books of enduring scholarly value (including out-of-copyright works originally issued by other publishers) across a wide range of disciplines in the humanities and social sciences and in science and technology.

Economics of Construction in Relation to Framed Structures

Robert H. Bow

CAMBRIDGE
UNIVERSITY PRESS

University Printing House, Cambridge, CB2 8BS, United Kingdom

Cambridge University Press is part of the University of Cambridge.
It furthers the University's mission by disseminating knowledge in the pursuit of education, learning and research at the highest international levels of excellence.

www.cambridge.org
Information on this title: www.cambridge.org/9781108071932

This edition first published 1873
This digitally printed version 2014

ISBN 978-1-108-07193-2 Paperback

ECONOMICS OF CONSTRUCTION

IN RELATION TO

FRAMED STRUCTURES

ECONOMICS OF CONSTRUCTION

IN RELATION TO

FRAMED STRUCTURES

BY ROBERT H. BOW, C.E., F.R.S.E.

PRELIMINARY PARTS.

PART I.—CLASSIFICATION OF STRUCTURES.

PART II.—DIAGRAMS OF FORCES.

LONDON: E. & F. N. SPON, 48 CHARING CROSS
NEW YORK: 446 BROOME STREET
EDINBURGH: ADAM & CHARLES BLACK
1873

Part II. gives the application of the recently expanded method of arriving at the stresses in structures by drawing diagrams of the forces. The importance to which this method has attained is almost altogether due to Professor J. Clerk Maxwell, who has shown how surprisingly general its application is, and who has placed it in quite a new aspect by his discovery or detection of those diagrams of forces which bear a reciprocal relationship to the relative framed structures.

When I first gave particular attention to the method, it was with a view to preparing an appendix to the general work I projected; but having devised an improvement in the mode of lettering the diagrams, I was led on to devote a considerable amount of time to the subject; and the notes and particulars of its application accumulating on my hands, I have thought them deserving of separate publication.

Some novelty may also be claimed for the application of the method to the imperfect structures constituting Class II.

7 South Gray Street, Edinburgh,
October 1873.

PREFACE.

THE two Parts now published are calculated to form a useful Manual in themselves. And as they are almost altogether of a preliminary character in relation to *Economics of Construction*, no introduction to that subject need here be attempted. But as touching upon the neglect of that branch of Engineering Science, I may quote the following passage with reference to British Engineering from *Applied Mechanics* by the late eminent Professor W. J. Macquorn Rankine :—" In too many cases we see the strength and the stability which ought to be given by the skilful arrangement of the parts of a structure, supplied by means of clumsy massiveness, and of lavish expenditure of material, labour, and money; and the evil is increased by a perversion of the public taste, which causes works to be admired, not in proportion to their fitness for their purposes, or to the skill evinced in attaining that fitness, but in proportion to their size and cost."

Part I. is devoted to the Classification of Structures, but the illustrations are, in the meantime, confined to Roof and Bridge Trusses. Many previous attempts have been made by others and myself in this direction, but not one of those that I am acquainted with is at all satisfactory. I trust that the classification now given, although no doubt open to some objections, will commend itself as based for the most part upon scientific distinctions.

A considerable number of the designs supplied to illustrate the classification are original.

ECONOMICS OF CONSTRUCTION

IN RELATION TO

FRAMED STRUCTURES.

PART I.

CLASSIFICATION

OF

FRAMED STRUCTURES OR TRUSSES.

NOTE.—In my Treatise on Bracing (1851), the term "Truss" was applied to any framed girder with a triangular outline, or which did not require any of the struts or ties to change their characters under irregular loadings; but this restriction of its meaning has not been adopted by subsequent writers, and now the term may be used to designate any form of framed structure employed to support a pressure or load over a void.

CLASSIFICATION OF TRUSSES.

DISTINCTIONS BETWEEN ROOF AND BRIDGE TRUSSES.

In roof trusses the line or surface on which the loading is chiefly imposed is, from its office of discharging rain and snow, necessarily a more or less inclined one. In ordinary bridge trusses, on the contrary, the line of the roadway at which the principal part of the loading is situated approximates to a perfect level. This important distinction leads to another very marked difference—a roof truss and its loading, if treated as *inverted*, has no practical value, whereas any good bridge truss becomes, when inverted, another useful design, the natures of the parts becoming simply reversed as regards their strut or tie action. Another distinction between roofs and bridges is that the former have to withstand important oblique forces arising from the action of the wind blowing in the direction of the plane of the truss against the inclined surfaces, whereas the action of the wind in the same plane in the case of bridge trusses is of no account.

TIED AND ABUTTING STRUCTURES.

An abutting truss is to be regarded as nearly in the same condition as the same structure supplied with a tie-bar and resting vertically upon the supports, the chief difference being that in the abutting arrangement the distance between the abutments, sometimes assumed theoretically as invariable, is practically liable to uncertain, and it may be important, changes; whereas, when the horizontal thrust or spreading tendency of the feet is resisted by a tie-bar, the change of span may be calculated with some certainty.

CLASSIFICATION OF FRAMED STRUCTURES OR TRUSSES.

Class I.

In this class are placed all framed structures or trusses which are theoretically perfect, or which are characterised by the following properties:—

1st. If at each joint the parts be supposed to be so connected together as to leave them free to rotate as though hinged, the frame will, nevertheless, be free from liability to distortion—that is, no such rotation at the joints can take place.

2d. Any one part of the structure may be made longer or shorter, without thereby inducing stresses.

3d. Slight irregularities in the lengths of the parts* arising from imperfect workmanship, elasticity, or other cause, have no practical effect upon the distribution of the stresses.

Class II.

includes those trusses which, if the joints were all regarded as hinged, could only maintain their shapes under equilibrating loadings, and which owe what powers they practically possess to resist too great a distortion when subjected to non-equilibrating loadings, to the stiffness of the joints, and to the transverse stiffness of the parts which become bent when distortion takes place. In such structures there is a deficiency of parts as compared with Class I.

Class III.

comprises trusses in which there is a redundance of parts. This redundancy renders the particular distribution of the

* This includes the *span* in abutting structures, liable to change from various causes.

stresses more or less dependent upon the peculiarities of the fitting, and upon the relative changes in the lengths of the parts, arising from elasticity, temperature, etc. The characteristic test of this class is that any one part cannot be made longer or shorter, without thereby inducing longitudinal stresses in other parts of the structure.

Class IV.

Into this Class may be put all other trusses. These have the defects of both the Classes II. and III.—that is, while they present a redundance of parts in one region, a deficiency is apparent at another, so that uncertain longitudinal stresses are introduced, and a non-equilibrating load makes a demand upon the transverse strength of some of the parts.

GROUPS, SUB-GROUPS, ARRANGEMENTS, AND MODIFICATIONS.

The designs for roofs and bridges embraced by any of these classes may be grouped according to the number of sub-spans or stretches into which the whole span is divided: and the trusses belonging to any group placed in sub-groups according to the number of polygons (usually triangles) contained *in* them. The sub-groups sometimes admit of subdivision into *arrangements* of the parts;* and these again into *modifications*, arising from variations in the inclinations of the members.

It is obvious that the importance and complexity of the truss will be roughly indicated by the number of its sub-spans; and similarly the complexity of the diagram of forces will be somewhat in proportion to the number of polygons in the truss.

* In the table of designs, the letters A, B, C, etc., indicating a succession of "Arrangements," are carried through each *group*, instead of sub-group, since many sub-groups contain each but one arrangement of parts.

Parts required merely to support the tie-bar, or give lateral stiffness to the long struts, may be occasionally indicated, but these parts as usually treated do not appear in the diagram of forces, and should not in strictness be recognised as affecting the number of the polygons in the truss.

On pages 17 to 39 will be found a collection of designs arranged according to this classification. Those for which diagrams of forces are afterwards given are distinguished by an asterisk, and the same reference numbers will be retained in the lithographed plates of Part II.

CONVERSION TO OTHER CLASSES OF SOME STRUCTURES BELONGING TO CLASS II.

When one or more quadrilateral openings occur in a truss belonging to Class II., as in Nos. 139, 149, 150, 157, etc., it may frequently be reduced to Class I. by subdividing one of the quadrilaterals into two triangles by means of a diagonal piece capable of acting in turn as a strut or tie, as the changes in the distribution of the loading or other external forces may require; but this addition may sometimes prove unsightly, as shown by Figs. 11, 17, and 19; as an alternative we may frequently divide the quadrilateral into three triangles, as in Figs. 13, 18, 30, 33, without introducing a redundance of parts. The most usual mode of treating it is, however, to supply two diagonal pieces, as in Figs. 167, 168, 169, 185, etc.; but this at once places the truss in Class III., for though it might be *assumed* that each piece could only act as a tie, or as a strut, and the one only when the other was inactive, yet practically a slight misfitting of the lengths of the pieces would render such assumptions incorrect, and even though accurately fitted, under one degree of severity of the stresses a change in the loading might undo the adjustment.

SOME TYPES OF DESIGNS ARE COMMON TO MANY GROUPS.

Some of the modes of arranging the parts of a truss are suitable for a variety of divisions of the span; but it will not be necessary to repeat each such type of design in the illustrated lists for every group to which it may be applicable. The following may be noted as some of the best examples of such general designs:—1*st.* Those in which a brace radiates from a central point in the truss to each otherwise unsupported point, as in Figs. 21 to 28, 63 to 66, 71 and 72, among roofs; and Figs. 205, 213, and 223, among bridges. 2*d.* Designs in which the braces, etc., radiate from two points in the truss, as in Nos. 40, 74, 106, 230, etc. And 3*d,* Those having the braces arranged in zig-zag fashion, and dividing the truss into the simplest chain of triangles, as in Figs. 68, 85, 98, 105, etc., among roofs; and Figs. 258, 259, 260, 327, etc., among bridges.

The merits of this last type of designs were little understood before the publication of my Treatise on Bracing, in January 1851, which was almost wholly devoted to their elucidation.

In its simplest form this arrangement of parts had been patented by Nash, Warren, etc., and in the compounded state of the lattice-bridge by Town in America; its more general application and the proper proportioning of the scantlings to the stresses were, however, strangely neglected. Indeed, it had been claimed as an advantage in Captain Warren's bridge that the structure could be built up of triangles all cast from the same pattern.

MODIFICATIONS OF ROOF-TRUSS ARRANGEMENTS.

The contour of the roof-surface, and the shape given to the tie-bar may be greatly varied, and the appearance of the design greatly changed, without causing any change in the relative positions of the parts. It is of course impracticable to give all the modifications of the arrangements which so arise. The series of figures, 21 to 28, 33 to 37, 39 to 45, 48 to 53, may be pointed out as exhibiting the extreme plasticity of some designs: in each of these series the designs are mere modifications of one arrangement of the parts.

The leading outlines of roofs are as follow:—the common isosceles triangle with slopes varying from a rise of 1 in $2\frac{1}{2}$ (Fig. 338, i.) to one of 1 in $1\frac{1}{2}$ (ii.) and sometimes much steeper when architectural effect is sought after (iii).

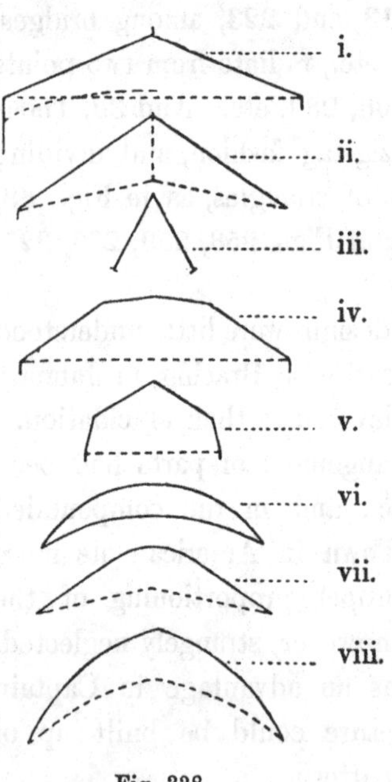

Fig. 338.

A shape becoming more common than formerly is that having the top more or less flattened (iv.) The central part may be covered in with zinc, or surmounted by the arrangements for lighting and ventilation.

Occasionally the upper part of the roof has the usual slopes, but with the lower parts made extremely steep (v.)

For some of the largest roofs an arc of a circle has been used for the form of the rafter, and the arc of a larger circle for the tie (vi.)

A form that has not yet been used, but which might offer some advantages over the circular, is the parabolic, or an approach thereto, by combining an arched crown with the ordinary sloped flanks (vii.)

The Gothic arch (viii.) might also be found good.

The part these different contours would play in an economic point of view must depend greatly upon the shapes given to the tie-bars. More will be said regarding this when we come to discuss some of the special designs.

Sometimes the end of the truss is finished, as shown at the left-hand side of Fig. 338 i., above, and in Fig. 190, so as to afford a rigid connection with the supporting column. This secondary office of giving rigidity to the whole erection will require additional material to meet the induced stresses.

MODIFICATIONS OF BRIDGE-FRAME ARRANGEMENTS.

Much that has been said regarding the modification of roof trusses might be repeated as applicable to framed bridges. In the tabulated designs for bridges few modifications are given, as the student may readily supply such. The leading modifications result from giving, triangular, quadrilateral, and segmented outlines to the truss, as in Figs. 232, 233, and 234; also the fish shape shown by Fig. 330. Modifications may also arise from variations in the inclinations of the interior parts of the truss, as in Figs. 275, 276.

RELATION BETWEEN THE NUMBER OF POLYGONS AND THE NUMBER OF PARTS IN THE TRUSS.

The only artificial part of the foregoing classification is the ranging the trusses in the order of the number of polygons. There exists, however, a curious relation between the number of polygons and the number of parts for each Class; thus, let p be the number of polygons, then in all perfect trusses, or those belonging to Class I., the number of parts (intersections being regarded as joints) is $= 2p + 1$. Whereas in Class II. the number of parts is $= 2p + 1 + d$, where d represents the least number of parts it would be necessary to add in order to convert the truss into one of Class I. In Class III., putting e for the least number of parts, it would be necessary to abstract to reduce the truss to Class I., the number of parts is $= 2p + 1 - e$. And in Class IV. the number of parts is $= 2p + 1 + d - e$.

Thus in Class II., for Fig. 140, $d = 1$, $\therefore$ the number of parts in Fig. 140 $= 2p + 2 = 8$. For Fig. 148 the deficiency $d = 2$, $\therefore$ number of parts $= 2p + 3 = 11$. In Class III., for Fig. 172 it would only be necessary to take out one of the central parts, therefore the excess $e = 1$, and accordingly the number of parts is $= 2p$, or 12. In Fig. 178 it would be necessary to remove two parts at least $\therefore$ $e = 2$, and the number of parts in the truss is $= 2p + 1 - 2 = 19$. In Class IV. there is at the same time deficiency and excess of parts; for Fig. 198, $d = 1$ and $e = 2$, $\therefore$ the number of parts $= 2p + 1 + 1 - 2 = 2p = 18$; in Fig. 199, $d = 2$, and $e = 1$ $\therefore$ number of parts $= 2p + 1 + 2 - 1 = 2p + 2 = 18$; in Fig. 200, $d = 1$ and $e = 1$ and $\therefore$ number of parts $= 2p + 1 = 21$. A truss belonging to Class I. must have $2p + 1$ parts, but a truss having that relation of parts to polygons may belong to Class IV.

The distinctions between the classes are clearly marked; but although the examples of Classes II. and III. always have the defects appertaining to the class, the degree in which the defect is likely to be present in a frame of Class III. varies greatly. Thus the lattice-girder No. 324 may be confidently expected to be very little affected by any ordinary misfitting of one of the end pieces; whereas structures of the abutting kind are not only likely to be misfitted to the span originally, but the length of the span between the abutments, which plays the part of one of the lines of the framing, may undergo after changes of serious amount.

CLASSIFICATION OF ROOF TRUSSES.

CLASS I.

Number of Sub-spans.	Number of Polygons in each Truss.	"Arrangements" of the Parts.	"Modifications" of the Arrangements.	Reference numbers of the Trusses.	Figs.
I.	0	A	*a*	1	
"	"	"	*b*	2	
II.	1	A	*a*	3*	
"	"	"	*b*	4*	
"	"	"	*c*	5*	
"	2	B	*a*	6	
"	"	"	*b*	7*	
"	"	"	*c*	8	
"	3	C	*a*	9*	
"	"	"	*b*	10*	
III.	2	A	*a*	11*	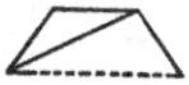
"	"	"	*b*	12	
"	3	B	*a*	13*	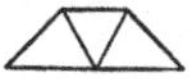
"	"	"	*b*	14*	
"	"	"	*c*	15	
"	"	"	*d*	16*	

* For all the trusses thus distinguished, diagrams of forces are given under the same numbers in the lithographed plates of Part II.

Remarks.—No. 2. See also Fig. 343. 3. When a tie-bar is indicated by a dotted line, as in this and some succeeding figures, the truss may be made either a tied or an abutting one. 4. See also Figs. 347 and 348. 8. See Fig. 350. 10. See also Figs. 349, 357, and 358.

CLASS I.—*Continued.*

Number of Sub-spans.	Number of Polygons in each Truss.	"Arrangements" of the Parts.	"Modifications" of the Arrangements.	Reference numbers of the Trusses.	Figs.
III.	4	C	*a*	17*	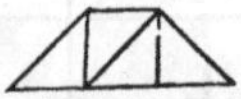
”	5	D	*a*	18*	
IV.	3	A	*a*	19*	
”	”	B	*a*	20*	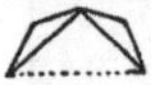
”	4	C	*a*	21*	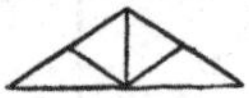
”	”	”	*b*	22*	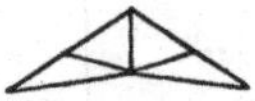
”	”	”	*c*	23*	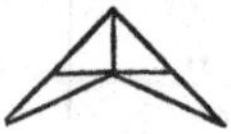
”	”	”	*d*	24*	
”	”	”	*e*	25*	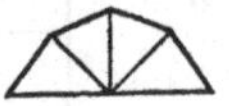
”	”	”	*f*	26*	
”	”	”	*g*	27*	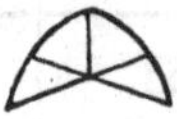
”	”	”	*h*	28*	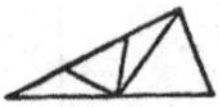
”	”	D	*a*	29*	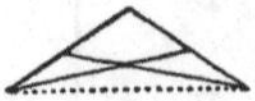
”	”	”	*b*	30*	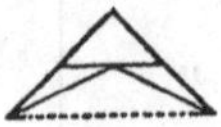
”	”	”	*c*	31	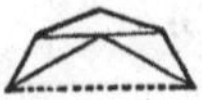
”	”	”	*d*	32*	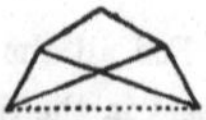

21 to 28. This set of designs is one of the most satisfactory for moderate spans, particularly in woodwork. Diagrams of forces are given for all the leading modifications of the Arrangement on Plates V. and VI.

32. In Plate IV. this is given as abutting, in Plate V. as tied.

CLASS I.—*Continued.*

Number of Sub-spans.	Number of Polygons in each Truss.	"Arrange-ments" of the Parts.	"Modifica-tions" of the Arrangements.	Reference numbers of the Trusses.	Figs.
IV.	4	E	*a*	33	
"	"	"	*b*	34*	
"	"	"	*c*	35	
"	"	"	*d*	36	
"	"	"	*e*	37	
"	"	F	*a*	38*	
"	5	G	*a*	39	
"	"	"	*b*	40*	
"	"	"	*c*	41	
"	"	"	*d*	42	
"	"	"	*e*	43*	
"	"	"	*f*	44*	
"	"	"	*g*	45	
"	"	H	*a*	46	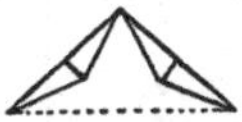
"	"	I	*a*	47	

40. See also Figs. 350 and 356.

CLASS I.—*Continued.*

Number of Sub-spans.	Number of Polygons in each Truss.	"Arrangements" of the Parts.	"Modifications" of the Arrangements.	Reference numbers of the Trusses.	Figs.
IV.	6	J	*a*	48	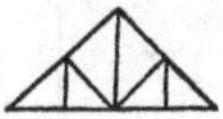
"	"	"	*b*	49*	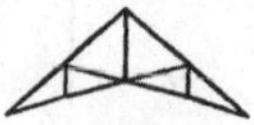
"	"	"	*c*	50*	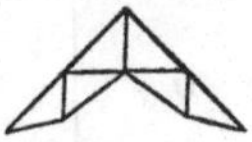
"	"	"	*d*	51*	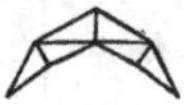
"	"	"	*e*	52	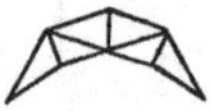
"	"	"	*f*	53	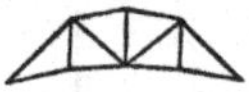
"	"	K	*a*	54*	
"	"	"	*b*	55	
"	"	L	*a*	56*	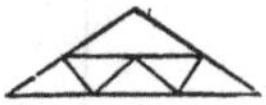
"	7	M	*a*	57	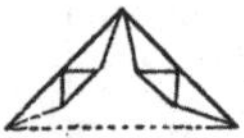
"	"	N	*a*	58	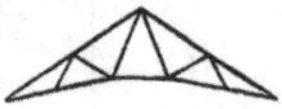
"	"	"	*b*	59*	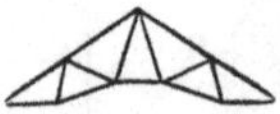
"	"	"	*c*	60	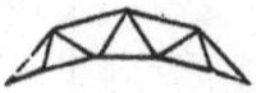
"	"	O	*a*	61*	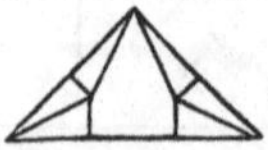
"	9	P	*a*	62	

CLASS I.—*Continued.*

Number of Sub-spans.	Number of Polygons in each Truss.	"Arrangements" of the Parts.	"Modifications" of the Arrangements.	Reference numbers of the Trusses.	Figs.
V.	5	A	*a*	63	
"	"	"	*b*	64*	
"	"	"	*c*	65	
"	"	"	*d*	66*	
"	6	B	*a*	67	
"	7	C	*a*	68*	
"	"	"	*b*	69	
"	"	"	*c*	70*	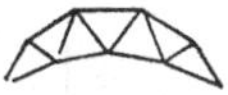
VI.	6	A	*a*	71*	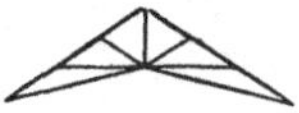
"	"	"	*b*	72	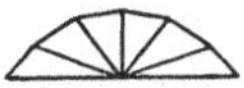
"	"	B	*a*	73*	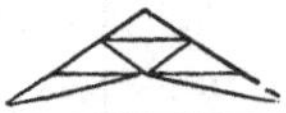
"	7	C	*a*	74*	
"	"	"	*b*	75	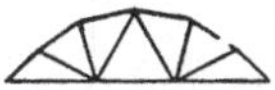
"	8	D	*a*	76*	
"	"	"	*b*	77	
"	"	"	*c*	78	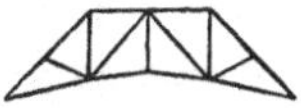
"	"	E	*a*	79	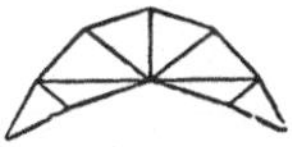

CLASS I.—*Continued.*

Number of Sub-spans.	Number of Polygons in each Truss.	" Arrange-ments " of the Parts.	" Modifica-tions " of the Arrangements.	Reference numbers of the Trusses.	Figs.
VI.	8	F	*a*	80*	
"	"	"	*b*	81	
"	"	"	*c*	82	
"	"	G	*a*	83	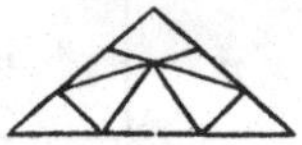
"	9	H	*a*	84*	
"	"	"	*b*	85	
"	"	I	*a*	86*	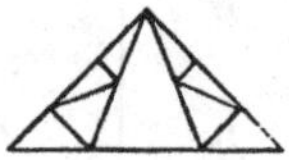
"	"	J	*a*	87*	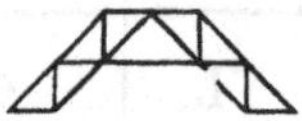
"	10	K	*a*	88*	
"	"	"	*b*	89*	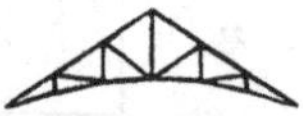
"	"	"	*c*	90*	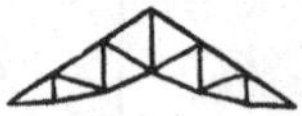
"	"	"	*d*	91	
"	"	L	*a*	92*	
"	"	"	*b*	93	
"	"	M	*a*	94*	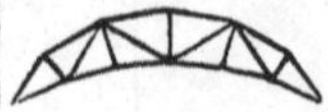

CLASS I.—*Continued.*

Number of Sub-spans.	Number of Polygons in each Truss.	"Arrangements" of the Parts.	"Modifications" of the Arrangements.	Reference numbers of the Trusses.	Figs.
VI.	10	N	*a*	95*	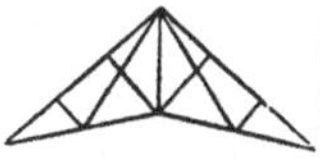
"	"	"	*b*	96	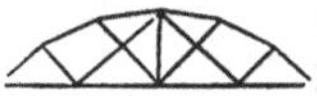
"	"	O	*a*	97	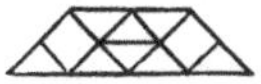
"	11	P	*a*	98*	
VII.	7	A	*a*	99	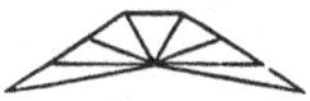
"	9	B	*a*	100*	
"	"	C	*a*	101*	
"	"	"	*b*	102	
"	10	D	*a*	103*	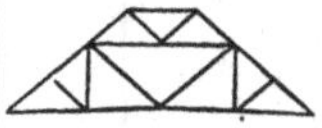
"	11	E	*a*	104*	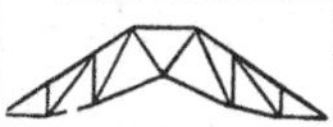
"	"	"	*b*	105*	
VIII.	9	A	*a*	106*	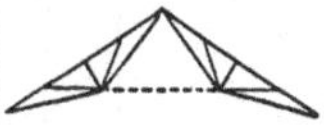
"	10	B	*a*	107*	
"	"	C	*a*	108*	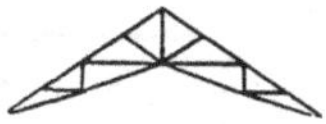
"	"	D	*a*	109	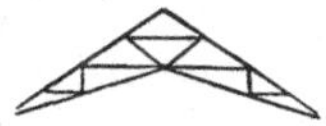

In Plate X. 106 is given in combination with 116, and 107 with 119. The roof of London Road Station, Manchester, is another modification of 108; it has a span of 51½ feet.

CLASS I.—*Continued.*

Number of Sub-spans.	Number of Polygons in each Truss.	"Arrangements" of the Parts.	"Modifications" of the Arrangements.	Reference numbers of the Trusses.	Figs.
VIII.	11	A	*a*	110*	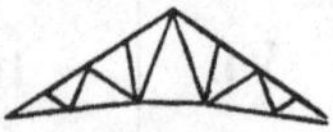
"	"	"	*b*	111*	
"	"	B	*a*	112	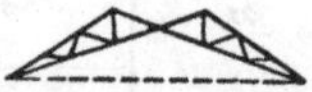
"	"	C	*a*	113	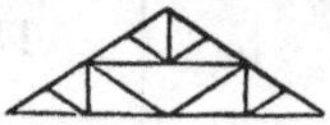
"	12	A	*a*	114*	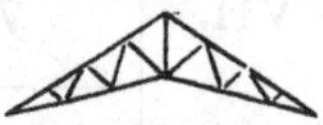
"	"	B	*a*	115	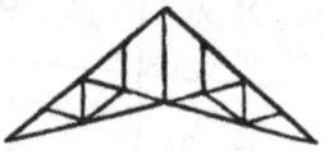
"	13	A	*a*	116*	
"	"	B	*a*	117*	
"	"	"	*b*	118*	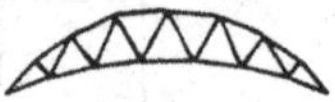
"	14		*a*	119*	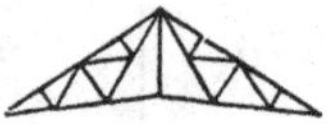
"	"	B	*a*	120	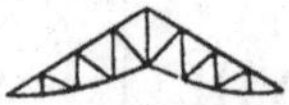
"	"	C	*a*	121	
IX.	11	A	*a*	122	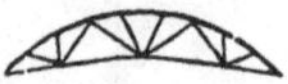
X.	14	A	*a*	123	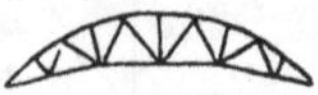
	15	A	*a*	124	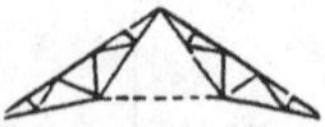

110 and 117. The diagrams for these are given in combination. 115. See No. 127.

CLASS I.—*Concluded.*

Number of Sub-spans.	Number of Polygons in each Truss.	"Arrangements" of the Parts.	"Modifications" of the Arrangements.	Reference numbers of the Trusses.	Figs.
XII.	15	A	*a*	125	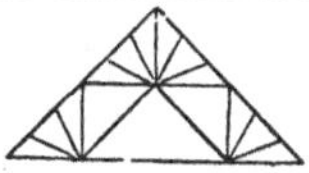
"	17	B	*a*	126*	
"	20	C	*a*	127*	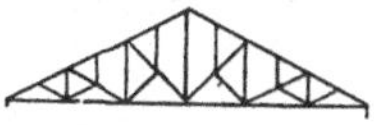
"	"	D	*a*	128*	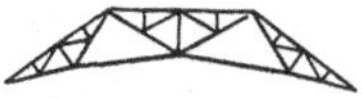
"	21	E	*a*	129*	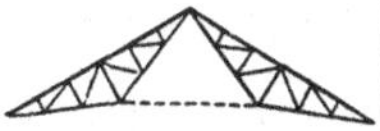
"	22	F	*a*	130	
Irregular, or Indefinite, or more numerous than required.		A	*a*	131	
		B	*a*	132*	
		C	*a*	133	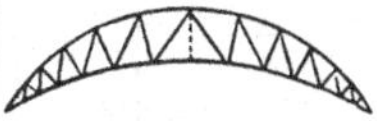
		D	*a*	134	
		E	*a*	135	
		F	*a*	136	
		G	*a*	137*	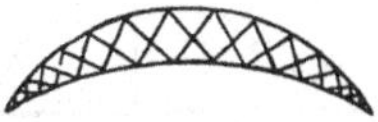

126. The diagram for this is given along with that for **129.**

127. Designed by Mr. Bow in 1853 : this is very suitable for wood-work.

130. Roof by Messrs. Fox and Henderson for a span of 86¼ feet.

137. Designed by Mr. Bow in 1871 for a span of 97½ feet.

Other designs of this class will readily suggest themselves. In No. **132** we have one series of braces, in **137** two series; a greater number may be employed ; or intermediate points of support to the rafter may be supplied by struts inserted in various manners in these and other forms of roofs.

CLASS II.

Number of Sub-spans.	Number of Polygons in each Truss.	"Arrangements" of the Parts.	"Modifications" of the Arrangements.	Reference numbers of the Trusses.	Figs.
III.	1	A	*a*	138	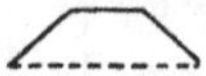
”	3	B	*a*	139*	
”	”	”	*b*	140	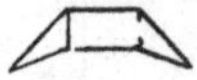
IV.	2	A	*a*	141*	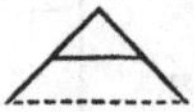
”	3	B	*a*	142	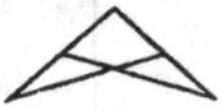
”	”	”	*b*	143	
”	”	”	*c*	144	
”	4	C	*a*	145*	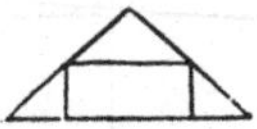
”	”	D	*a*	146	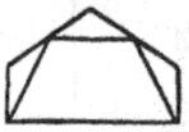
V.	2	A	*a*	147	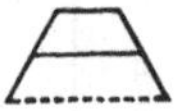
”	4	B	*a*	148	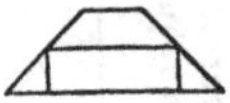
”	5	C	*a*	149	
”	6	D	*a*	150	
”	”	”	*b*	151	
VI.	4	A	*a*	152	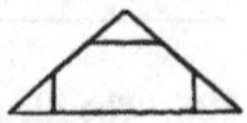
”	6	B	*a*	153	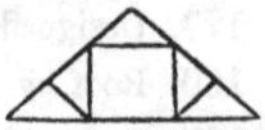

142 is very imperfect.

139, 145, 149, 153. These are practically good in woodwork when a stiff tie-beam is used.

CLASS II.—*Continued.*

Number of Sub-spans.	Number of Polygons in each Truss.	"Arrangements" of the Parts.	"Modifications" of the Arrangements.	Reference numbers of the Trusses.	Figs.
VI.	7	C	*a*	154	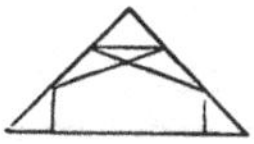
"	9	D	*a*	155	
"	"	E	*a*	156	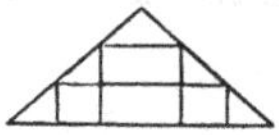
"	"	F	*a*	157	
"	"	"	*b*	158	
"	"	G	*a*	159	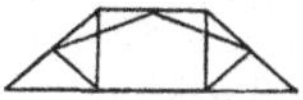
VII.	8	A	*a*	160	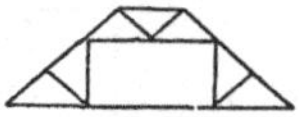
VIII.	8	A	*a*	161	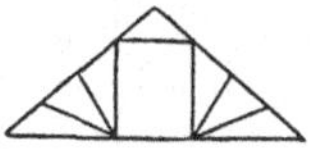
"	9	B	*a*	162	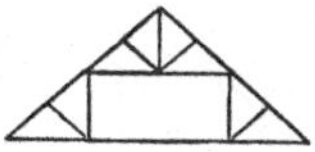
X.	13	A	*a*	163	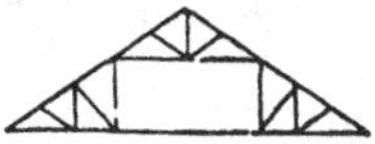
XII.	13	A	*a*	164	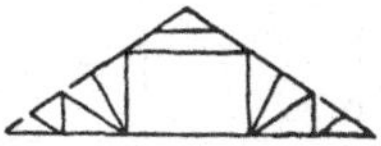
				165	 Ship-building Shed.

CLASS III.

Number of Sub-spans.	Number of Polygons in each Truss.	"Arrangements" of the Parts.	"Modifications" of the Arrangements.	Reference numbers of the Trusses.	Figs.
II.	3	A	*a*	166*	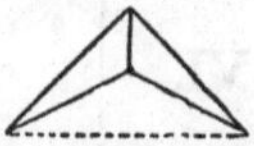
III.	4	A	*a*	167	
"	6	B	*a*	168	
IV.	5	A	*a*	169	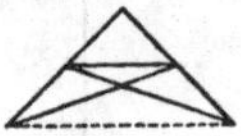
"	"	B	*a*	170	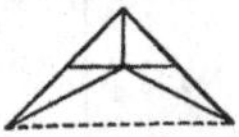
"	"	"	*b*	171	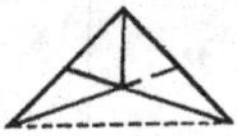
"	6	C	*a*	172	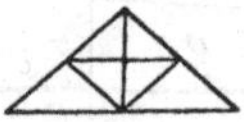
"	7	D	*a*	173	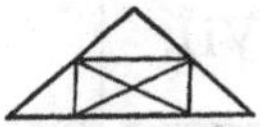
"	"	E	*a*	174	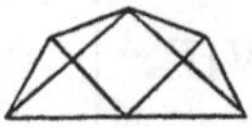
"	8	F	*a*	175	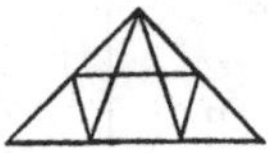
"	9	G	*a*	176	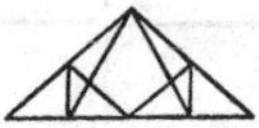
"	"	"	*b*	177	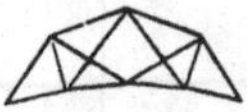
"	10	H	*a*	178	
V.	6	A	*a*	179	

CLASS III.—*Continued.*

Number of Sub-spans.	Number of Polygons in each Truss.	"Arrangements" of the Parts.	"Modifications" of the Arrangements.	Reference numbers of the Trusses.	Figs.
V.	8	B	*a*	180	
"	"	"	*b*	181	
"	"	C	*a*	182	
"	10	D	*a*	183	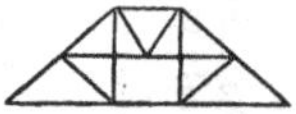
"	"	"	*b*	184	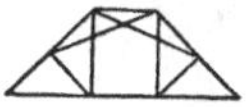
"	12 or 14	E or F	*a*	185	
VI.	9	A	*a*	186	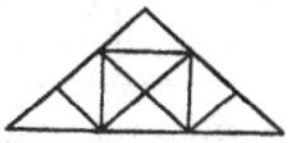
"	11 or 12	B or C	*a*	187	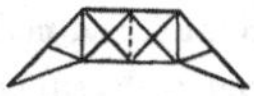
VIII.	12	A	*a*	188	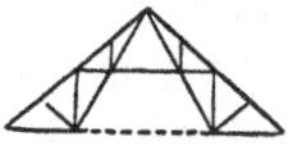
"	15	B	*a*	189	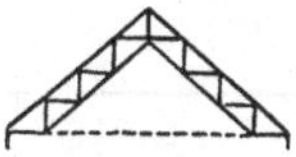
"	17	C	*a*	190	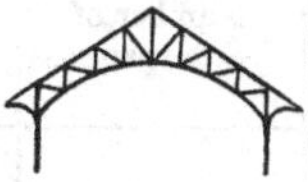
IX.	18	A	*a*	191	

190 is here assumed to be abutting.

CLASS III.—*Concluded.*

Sub-spans.	Arrangements.	Remarks.	Reference numbers of the Trusses.	Figs.
Subdivisions of the Span, irregular in size or indefinite in number.	A	Roof over Lime Street Station, Liverpool. Span about 133 feet.	192	
	B	...	193	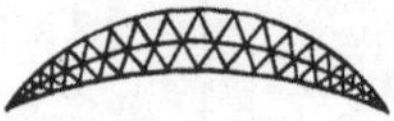
	C	...	194	
	D	Roof over Joint Railway Station, Birmingham, 1853. Span 211 ft.	195	
	E	Abutting or Bracket action.	196	
	F	Abutting	197	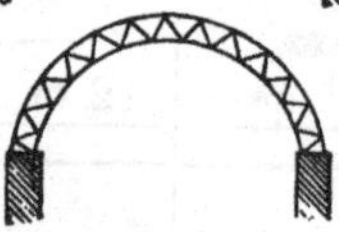

Class III. also includes all the structures of Class I. of the character of Nos. **51**, **70**, **87**, **93**, **137**, and many others, when these are made abutting, or supplied with tie-bars on the level of the supports.

CLASS IV.

Number of Sub-spans.	Number of Polygons.	Arrangements.	Reference numbers of the Trusses.	Figs.
III.	9	A	198	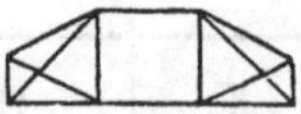
V.	8	A	199	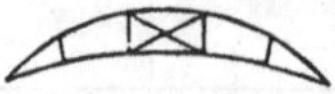
”	10	B	200	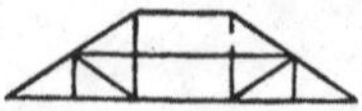
VI.	13	A	200½	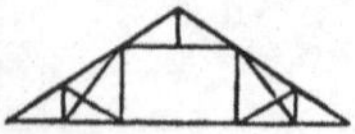

CLASSIFICATION OF FRAMED BRIDGES.

CLASS I.

			Remarks.	Figs. and References Numbers of the Trusses or Designs.
No. of Sub-spans = II.	No. of Polygons = 2.	Arrangement A.	Fig. 201 iii. shows No. 201 i. in the inverted state.	201* i. 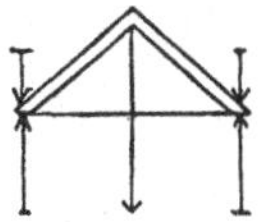
			N.B.—Only one drawing will be given for each of the succeeding designs, but they are all suited to be used in the inverted forms.	201* iii. 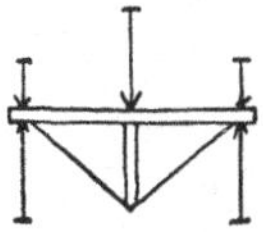
	No. of Polygons = 3.	Arrangement B.	202 A. The loading is here shown as supported by the "Bracket Action" of each half of the structure.	202* A 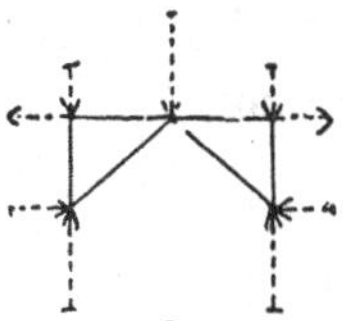
			202 B. This is the same form of structure as the above, but the loading is here shown as supported through "Abutting Action."	202* B 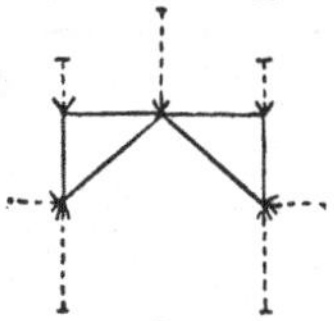
			202 C. Same truss as the above, but supplied with a tie-bar.	202* C
			N.B.—In the Figures which follow, the Abutting and Tied states will not be given separately; a dotted line for the tie-bar will generally be shown, to indicate that the tie may be used or not. But in counting the polygons, the truss will be considered as tied.	

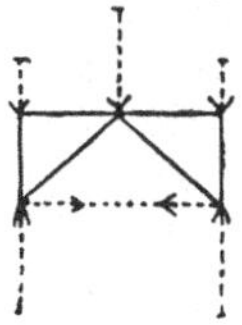

* Diagrams of Forces are afterwards given for all the designs marked with an asterisk.

CLASS I.—*Continued.*

Number of Sub-spans.	Number of Polygons in each Truss.	"Arrangements" of the Parts.	"Modifications" of the Arrangements.	Reference numbers of the Trusses.	Figs.
II.	3	C	*a*	203	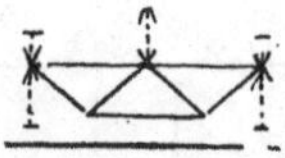
III.	3	A	*a*	204	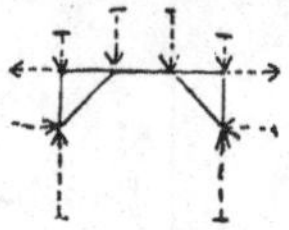
"	"	B	*a*	205*	
"	4	C	*a*	206	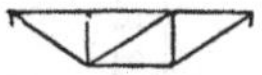
"	5	D	*a*	207	
"	"	"	*b*	208*	
"	"	"	*c*	209	
"	"	E	*a*	210	
"	"	F	*a*	211*	
"	7	G	*a*	212	
IV.	4	A	*a*	213*	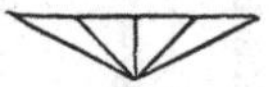
"	5	B	*a*	214	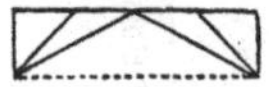
"	"	C	*a*	215	
"	6	D	*a*	216	
"	"	"	*b*	217*	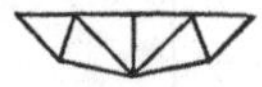
"	"	E	*a*	218	
"	"	F	*a*	219*	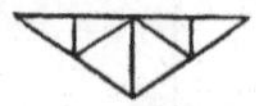
"	7	G	*a*	220*	
"	"	"	*b*	221	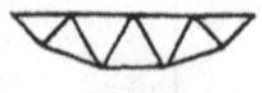
"	8	H	*a*	222*	
V.	5	A	*a*	223	

204 is the same in appearance as 278 of Class II.; but when the terminal triangles can act effectively as brackets, as shown here, the truss belongs to Class I.

CLASS I.—*Continued.*

Number of Sub-spans.	Number of Polygons in each Truss.	"Arrangements" of the Parts.	"Modifications" of the Arrangements.	Reference numbers of the Trusses.	Figs.
V.	7	B	*a*	224	
,,	,,	,,	*b*	225*	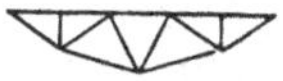
,,	,,	,,	*c*	226	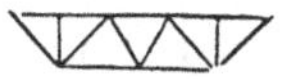
,,	9	C	*a*	227	
,,	,,	,,	*b*	228*	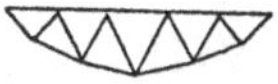
VI.	7	A	*a*	229	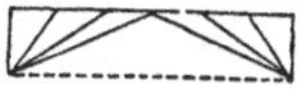
,,	,,	B	*a*	230*	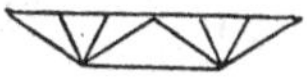
,,	8	C	*a*	231*	
,,	,,	D	*a*	232	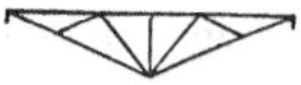
,,	,,	,,	*b*	233	
,,	,,	,,	*c*	234	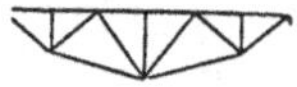
,,	9	E	*a*	235	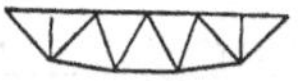
,,	,,	,,	*b*	236	
,,	,,	F	*a*	237	
,,	10	G	*a*	238	
,,	,,	H	*a*	239	
,,	11	I	*a*	240	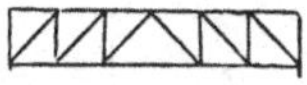
,,	,,	J	*a*	241	
,,	,,	,,	*b*	242	
,,	12	K	*a*	243	
,,	14	L	*a*	244*	

CLASS I.—*Continued.*

Number of Sub-spans.	Number of Polygons in each Truss.	"Arrangements" of the Parts.	"Modifications" of the Arrangements.	Reference numbers of the Trusses.	Figs.
VII.	9	A	*a*	245	
„	10	B	*a*	246‡	
„	11	C	*a*	247	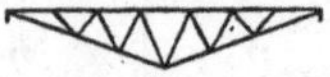
„	13	D	*a*	248	
VIII.	9	A	*a*	249	
„	10	B	*a*	250	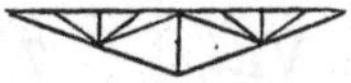
„	11	C	*a*	251	
„	„	D	*a*	252	
„	12	E	*a*	253	
„	„	F	*a*	254‡	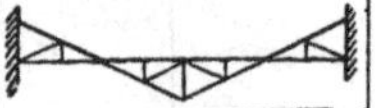
„	13	G	*a*	255*	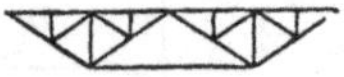
„	„	H	*a*	255½*	See Plate XIV.
„	14	I	*a*	256	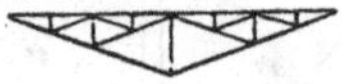
„	„	J	*a*	257	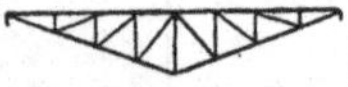
„	„	K	*a*	258	
„	15	L	*a*	259	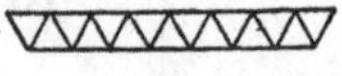
„	„	„	*b*	260	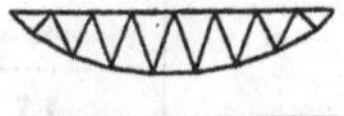
„	„	M	*a*	261	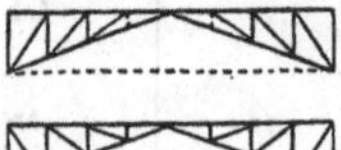
„	„	N	*a*	262	
„	„	O	*a*	263	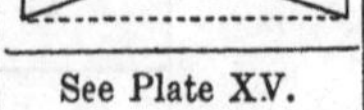
„	17	P	*a*	263½*	See Plate XV.

‡ 246 and 254 are supposed to be supported by the efficient bracket-action of the portions adjoining the piers.

CLASS I.—*Concluded.*

Note.—A great variety of designs may be produced by assuming the existence of efficient bracket-action, as in Figs. 202 A, 204, 246, and 254; further examples need not be given.

Number of Sub-spans.	Number of Polygons in each Truss.	"Arrangements" of the Parts.	"Modifications" of the Arrangements.	Reference numbers of the Trusses.	Figs.
IX.	11	A		264	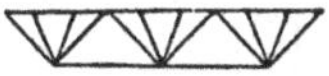
"	17	B		265	
"	19	C		266‡	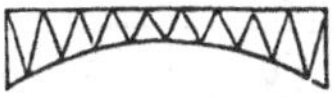
"	20	D		267*	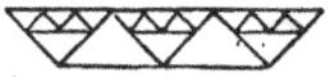
"	"	E		268	
X.	14	A		269	
"	15	B		270	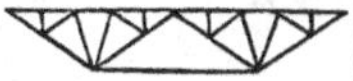
"	17	C		271	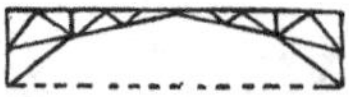
"	19	D		272	
"	"	E		273	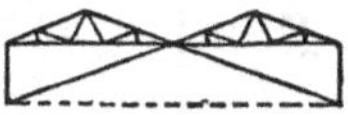
XII.	23	A		274	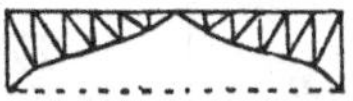
"	29 or 31	B or C	*a*	275	
"	31	C	*b*	276	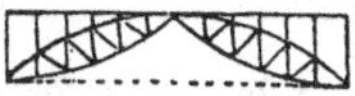

See also Fig. 37 in Treatise on Bracing.

Examples need not be here given of trusses having the subdivisions of the span irregular in size or indefinite in number.

‡ See 263½ on Plate XV.

CLASS II.

Number of Sub-spans.	Number of Polygons in each Truss.	" Arrangements " of the Parts.	Reference numbers of the Trusses.	Figs.
II.	3	A	277 ‡	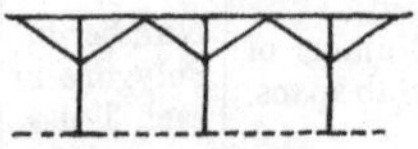
III.	3	A	278	
"	"	B	279	
IV.	4	A	280	
V.	5	A	281	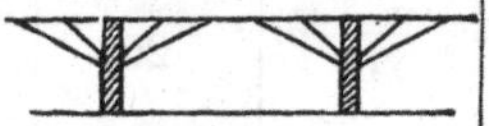
"	10	"	282	
VI.	11	A	283 ‡	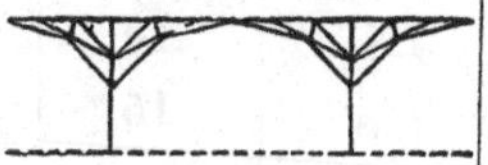
VII.	"	A	284	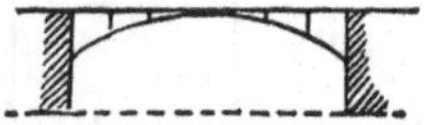
X.	17	A	285	
"	17	B	286	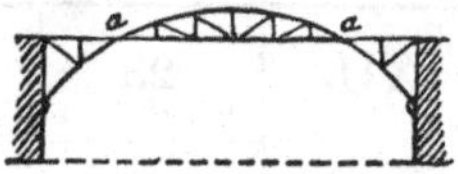
Indefinite.	Indefinite.	A	287	

‡ In 277 and 283 the abutments or piers are supposed to be flexible, and the stability of the structures dependent upon the resistance of these to extreme bending.

In Figs. 277, 283, 284, 285, abutting or bracket action is supposed though not specially indicated.

278, 281, 286. No efficient bracket-action is supposed to have place in these, otherwise they would belong to Class I.

Many other examples of this class might be drawn from existing structures.

CLASS III.

Number of Sub-spans.	Number of Polygons in each Truss.	"Arrangements" of the Parts.	"Modifications" of the Arrangements.	Reference numbers of the Trusses.	Figs.
II.	4	A	A great variety of "Modifications" of the arrangements shown on this and other pages might be drawn.	288	
"	7	B		289	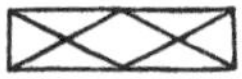
"	8	C		290	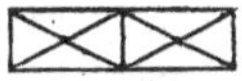
III.	6	A		291	
"	10	B		292	
"	10 or 12	C or D		293	
IV.	7	A		294	
"	8	B		295	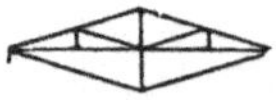
"	9	C		296	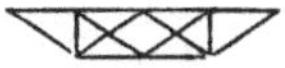
"	"	D		297	
"	10	E		298	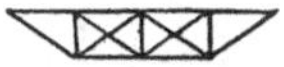
"	"	F		299	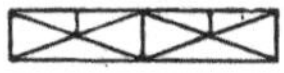
"	11	G		300	
"	13	H		301	
"	16	I		302	
V.	8	A		303	
"	10	B		304	
"	12	C		305	
"	"	D		306	
"	15	E		307	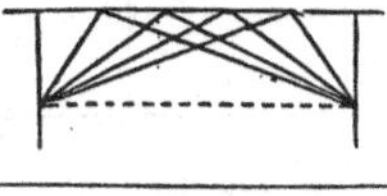

No. 293. If the dotted lines be omitted this is the too simple truss employed by Brunel for so great a span as 296 feet in his bridge at Chepstow.

CLASS III.—*Continued.*

Number of Sub-spans.	Number of Polygons in each Truss.	"Arrangements" of the Parts.	"Modifications" of the Arrangements.	Reference numbers of the Trusses.	Figs.
VI.	9	A		308	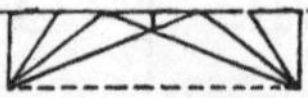
„	11	B		309	
„	13	C		310	
„	14	D		311	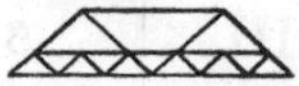
„	15	E		312	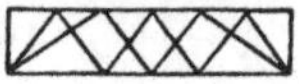
„	16	F		313	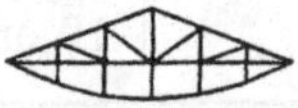
„	17	G	*a*	314	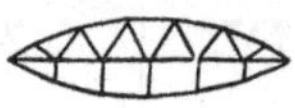
„	„	„	*b*	315	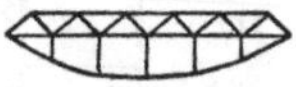
„	18	H		316	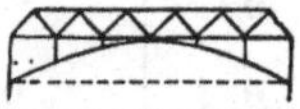
VII.	20	A		317	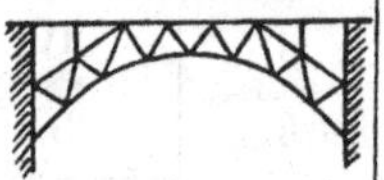
„	22	B		318	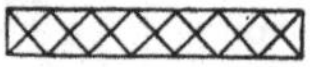
VIII.	14 or 20	A and B		319	
„	15	C		320	
„	„	D		321	
„	16	E		322	
„	17	F		323	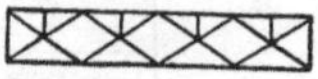
„	25	G		324	
„	26	H		325	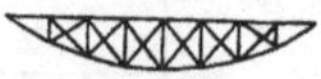
„	32	I		326	

308. This, in its inverted form, is one variety of Mr. Ordish's patent suspension bridges.

CLASS III.—*Concluded.*

Number of Sub-spans.	Number of Polygons in each Truss.	Remarks.	Reference numbers of the Trusses.	Figs.
IX.	20	Figs. 39 & 65 of Treatise.	327	
X.	32	See Figs. 43 & 66 of do.	328	
”	38	See Fig. 32 of do.		
XII.	30		329	
”	54?	Viaduct at Saltash, by Mr. Brunel.	330	
XVI.	84		331	
XXIV.	41	From drawings prepared by Mr. Bow, for Mr. T. Bouch, for a Viaduct (span of 204 feet) proposed to be carried over the Tyne at Newcastle.	332	

324 is the leading design of the viaducts and bridges on the South Durham and Lancashire Union Railway. Mr. Bow assisted Mr. Bouch in getting out the general and detail drawings for these. The depth is made one-eighth of the span, a proportion advocated by Mr. Bow in 1855. See *The Civil Engineer and Architect's Journal*, vol. xviii. page 236.

The Note at the end of Class III. of Roofs is equally applicable here. Thus Fig. 266 of Class I. becomes 327 of Class III. when made abutting, or supplied with a tie-bar.

Several forms of Girders which belong to Class I., if used simply for one span, become, when made continuous over piers, structures of the third class. The amount of bracket-action so introduced is affected by the fitting together, and may therefore be to some extent regulated on a method analogous to that proposed for controlling the position of the line of pressures in No. 327.

CLASS IV.

Number of sub-spans.	Number of Polygons in each Truss.	"Arrangements" of the Parts.	Reference numbers of the Trusses.	Figs.
VI.	9	A	333	
VII.	13	A	334	
IX.	21	A Abutting	335	
"	25	B Abutting	336	
"	35	C Abutting	337	

The designs for roofs and bridges may be multiplied indefinitely by subdividing the sub-spans of the simpler forms according to various methods, many of which are shown in the foregoing examples. The method of subdivision indicated by Figs. 272 or 273 is peculiar in not introducing complication into the stresses in the parts of the original framing, the loaded platform excepted.

CLASSIFICATION OF STRUCTURES.

In the foregoing examples the parts composing the structure, and the lines of action of the external forces, have been regarded as all lying in one plane—the plane of the paper. The principle of the classification is, however, applicable to all framed structures. I shall give here only one exemplification of structures of three dimensions, the case being that of single weight supported by props.

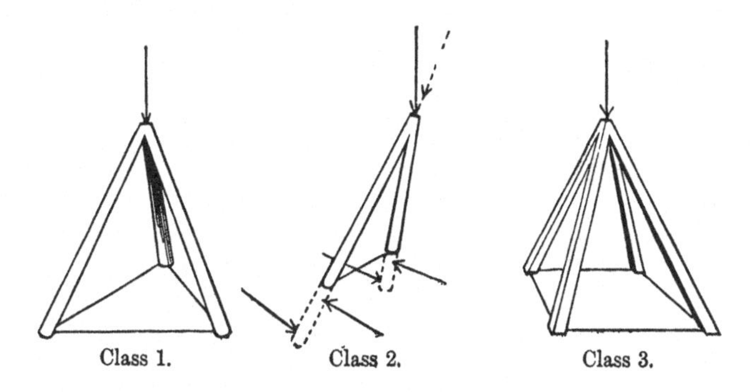

Class 1. Class 2. Class 3.

In the three-limbed support of Class I. there can be no dubiety about the amount of the stress each prop has to withstand.

But the two props constituting the support in Class II. could only be equilibrated by a force whose line of action lay in the plane of the props as shown by the dotted arrow; to resist the vertical force indicated by the black arrow, a demand is made upon the transverse strengths of the props.

The example of Class III. will give a clear idea of the effects of misfitting. *Theoretically* it might be assumed that each prop supported its full share of the load, but in reality it would be impracticable to cause the distribution of the stresses to approach with certainty, even roughly, to such equality; and then a very slight failure of the foundation of one of the props would not only relieve it from stress, but also the prop diagonally opposite to it; so that the other two props would each have to withstand a greater stress than if there had originally been only three, as in Class I.

ECONOMICS OF CONSTRUCTION

IN RELATION TO

FRAMED STRUCTURES

PART II.

ON

DIAGRAMS OF FORCES

WITH

NUMEROUS EXAMPLES.

DIAGRAMS OF THE APPLIED FORCES AND OF THE FORCES OF REACTION OF THE PARTS SUBJECTED TO STRESS IN FRAMED STRUCTURES.

USES OF THE METHOD.

DIAGRAMS of forces must render great assistance to any one occupied in designing roofs, bridges, or other framed structures. They furnish indeed a royal road to certain results, but this power of affording help to the helpless must not be regarded as their legitimate office. By means of a diagram of forces for any arrangement of framework provided for him, a person, quite incapable otherwise of taking out the stresses in a moderately complicated structure, could get at these for any particular modification of the arrangement with some assurance of accuracy; but such a one could never properly be entrusted with the duty of choosing the design, of making the detail drawings, or of superintending the erection of the work. In the operation of drawing details judiciously, there is a constant necessity for a reasoning appreciation of the various stresses and strains* that may act in the parts, as well as a practical acquaintance with the nature of materials, their strengths and their imperfections—acquirements that should place "Detailing" second only among professional duties to that of determining the most economical or best design to meet the occasion.

The uses of diagrams of forces are—not to render it unnecessary that the engineer or architect should familiarize himself with the mechanical interaction of the parts of framed

* The term *stress* expresses the condition of a part of the structure to the extremities of which are applied compressing or extending forces; the amount of the stress is measured by the magnitude of the force acting on either extremity; the *strain* is the change of length from elasticity which the part undergoes when subjected to the stress.

and other structures—but to facilitate the work of taking out the stresses, to suggest rapidly improvements in the design, and to reduce the labour of comparing the economic merits of different designs.

EARLIER EXAMPLES OF DIAGRAMS OF FORCES.

The representation of interacting forces by a diagram has been long practised in its simpler forms to determine the magnitudes of the stresses at particular points or parts of framed structures. Fig. 341 i. (Plate I.) shows an application of the method that must be familiar to any one accustomed to such work; in it the line B C is drawn vertically, and so that its length will represent on a chosen scale the magnitude of the vertical reaction of the support A′A (any weight A D coming directly upon the support being neglected); then the stresses in the extreme parts of the rafter and tie-bar are at once ascertainable by measuring on the same scale the respective lengths of A B and A C, cut off by B C. This gives us one of the simplest examples of a diagram of forces. It is reproduced in the form of two separate drawings in Figs. 341 ii. and iii.; ii. representing the end of the truss with the supporting force A′A acting upon it, and iii. the diagram of the forces. In ii. we have the three forces acting upon a *point*, while in iii. the lines parallel to and representing in magnitude these forces become the sides of a *triangle.* If A′A be oblique instead of vertical, then B C in the diagram must of course be drawn parallel to it.

Fig. 342 i. is an example taken from a paper by the late Mr. C. H. Wild, C.E., which was shown to me in the year 1854, but was of much earlier date. With the frame and diagram given separately, this is shown by Figs. 342 ii. and iii.; the stress in any part in ii. is measured by the length of the corresponding line in iii.

The next example of a "diagram of forces," shown by Fig.

343, is taken from Professor Rankine's *Applied Mechanics*, 1858, section 142. In this the three forces act upon a solid or frame A B, and in order that equilibrium may be produced, it is necessary that the lines of action of the forces shall intersect in some one point C; the triangle, Fig. 343 ii., formed by drawing lines parallel to the forces, constitutes the diagram of forces—that is, the lengths of the lines in it give the proportionate magnitudes of the forces to which they are respectively drawn parallel.

The well-known diagram of forces for the funicular polygon is shown by Fig. 344 ii. Let A B C D E in Fig. 344 i. represent a string the weight of which may be neglected; let it be suspended from the points A and E at the same level, and let any number of weights, w_1, w_2, w_3, be hung from it as in the figure. Then the relative magnitudes of the weights which cause the string to assume the particular form shown, and of the resulting stresses in the various parts of the string, can be readily ascertained by drawing the diagram of forces Fig. 344 ii.; to construct this, we must draw from the same point F lines parallel to the different parts of the string, and intersect these by a vertical line G J. Now if the length of F J be taken to represent the stress in the part A B to which it is parallel, J K will measure its vertical component (F K being drawn horizontally); similarly the length of F I will give the stress in the part B C, and I K its vertical component, and the difference between these vertical components = I J will represent the weight w_1; in like manner H I = w_2 and G H = w; and the stress in C D is measured by F H, and that in D E by F G. It is obvious that the horizontal component of the stress at any part of the string must be the same, and it is measured by F K.

If we suppose the funicular polygon to become inverted, and its parts capable of acting as struts, we get the polygonal frame Fig. 345 i., which will be equilibrated by the external forces given by the diagram ii. constructed in the same manner as the previous one; and if we assume the frame to be tied, J K measures the supporting force A′A, J F = compression in B A, J I = w_1, and so on, F K being the horizontal component of the thrusts in B A, etc., or the tension in A E.

Professor Rankine, in his *Applied Mechanics*, at section 150, has given the *diagram* of *forces* for the polygonal frame equilibrated by forces acting in *any* direction, and this must be regarded as a very important stride forward in the development of the method. In the diagram 345 ii. we see that J I represents w_1, I H = w_2, and H G = w_3; at the same time J K = A′A, and G K = E′E, or we may make this clearer by representing the diagram slightly distorted, as in Fig. 345 iii. Now Professor Rankine pointed out that instead of drawing all the lines representing the external forces as parts of one vertical line, as seen in 345 ii., we may draw them in any directions provided we form with them a closed polygon, such as G H I J K G in Fig. 346 ii.; and then a frame as 346 i. composed of parts parallel to the lines radiating from any point F in 346 ii. will be equilibrated under forces whose lines of action are parallel to the sides of the polygon, and of magnitudes proportional to the lengths of the same, as shown in the figure. At the same time the lengths of the lines radiating from F in ii. express the magnitudes of the stresses in the corresponding parts of the frame i.

At section 148, Professor Rankine has further given this simple but comprehensive case: let the sides of any triangle, A B C Fig. 347 i. or 348 i., represent in direction and magnitude three forces, and from any point F let lines be drawn to the angles A, B, and C; then a frame whose sides are parallel to F A, F B, and F C, will be equilibrated by three forces parallel and equal to A B, B C, and A C, as shown in Fig. 347 ii. or 348 ii.; and the stresses in the sides of the frame will be measured by the lengths of the lines parallel to them in the diagram.

At section 155 of *Applied Mechanics*, Professor Rankine gives a diagram of forces for a roof truss (see Fig. 349); and at section 119 of his *Civil Engineering* (1862), he gives the further example shown by Fig. 350 (with different lettering).

PARTICULARISATION OF A FORCE.

Five particulars regarding a force acting upon a body are required in order that it may be fully known—

1st. The "point of application," as point *a* in Fig. 351.

2d. The line in which the force acts, or the "line of action," as *b c* or *c b*.

3d. The direction in which the force acts in the "line of action," that is, whether it tends to move the point *a* towards *b* or towards *c*; this may be indicated by an arrow as shown in the figure.

4th. The side on which the other body lies, between which and point *a* the force acts, that is, whether the force acts as a push in *c a*, or as a pull in *a b*.

5th. The "magnitude" of the force.

All these particulars may be indicated by arrows as in Fig. 352, the length of the arrow being taken as a measure of the magnitude of the force.

It is usual to employ the word "Direction" to indicate the 2d, 3d, or 4th particulars, or any combination of these, and some confusion occasionally arises from the consequent indefiniteness. But I shall not here attempt to introduce new terms, trusting that when the word is used the context will indicate the intended meaning.

Let three forces be applied to the point *s*, Fig. 353, so as to be in equilibrium, and with lines of action parallel to A B,

B C, and C A:—required the magnitudes and other particulars of these forces.

By constructing the triangle Fig. 354, with sides drawn parallel to the three given lines, the lengths of these sides give us at once the relative magnitudes of the corresponding forces; but we have yet only the 1st and 2d particulars of each; if, however, the absolute magnitude of one, say A B, be given, the 5th particular becomes known for all. Again, if the 3d particular be given with respect to one of the forces, say that it is indicated by the arrow against the side A B, we arrive at the same particular concerning the other two forces by placing arrows against the sides in such a manner that they will all appear to travel round the triangle in the same direction. We have now arrived at all the particulars, except the 4th, for each of the three forces, but the 4th cannot be shown by or derived from the diagram of forces. Fig. 355 shows the various modes, eight in number, in which the three forces represented by diagram 354 may act on the point; and the 4th particular is required for each force before we can say which arrangement represents the case under investigation.* Generally, however, there is little difficulty in determining the 4th particular for each force from the nature of the structure.

DIAGRAMS OF FORCES: EXTENSION OF THE METHOD.

The method of exhibiting the forces acting in the parts of, and externally upon, a structure by means of a diagram, has been very fully investigated, and its application vastly extended by Professor Clerk Maxwell, who has at the same time shown that the diagrams may be so drawn that there shall be a striking and beautiful reciprocal relationship between them and the frames.

* Were the 3d particular not given, the number of modes in which the hree forces could act in equilibrium against the point would be 16.

The following extract from Professor Clerk Maxwell's Paper on "Reciprocal Figures, Frames, and Diagrams of Forces," given in the *Transactions* of the Royal Society of Edinburgh for Session 1869-70, supplies an account of the development of the method up to the date of the paper :—

"The properties of the 'triangle' and 'polygon' of forces have been long known, and a 'diagram' of forces has been used in the case of the 'funicular polygon,' but I am not aware of any more general statement of the method of drawing diagrams of forces before Professor Rankine applied it to frames, roofs, etc., in his *Applied Mechanics*, p. 137, etc. The 'polyhedron of forces,' or the proposition that forces acting on a point perpendicular and proportional to the areas of the faces of a polyhedron are in equilibrium, has, I believe, been enunciated independently at various times, but the application of this principle to the construction of a diagram of forces in three dimensions was first made by Professor Rankine in the *Philosophical Magazine*, February 1864. In the *Philosophical Magazine* for April 1864 I stated some of the properties of reciprocal figures, and the conditions of their existence, and showed that any plane rectilinear figure, which is a perspective representation of a closed polyhedron with plane faces, has a reciprocal figure. In September 1867 I communicated to the British Association a method of drawing the reciprocal figure, founded on the theory of reciprocal polars.

"I have since found that the construction of diagrams of forces, in which each force is represented by one line, had been independently discovered by Mr. W. P. Taylor, and had been used by him as a practical method of determining the forces acting in frames for several years before I had taught it in King's College, or even studied it myself. I understand that he is preparing a statement of the application of the method to various kinds of structures in detail, so that it can be made use of by any one who is able to draw one line parallel to another.

"Professor Fleeming Jenkin, in a paper recently published by the Society, has fully explained the application of the method to the most important cases occurring in practice."

DIAGRAMS OF FORCES, AND THE "RECIPROCAL" DIAGRAM OF FORCES.

For each frame we can draw a variety of useful diagrams of forces. To illustrate this, seven varieties for the frame Fig. 356 i. are given, under the numbers ii. to viii.: one of these—viz. 356 ii., bears certain reciprocal relationships to the frame and the external forces acting upon it, and this reciprocity being more complete for ii. than for any other possible diagram, it is distinguished as the "Reciprocal Figure," or "Reciprocal Diagram of Forces."

The chief reciprocal relationship is generally stated thus—"Corresponding lines which meet in a point in the one figure form a closed polygon in the other." This is, however, too comprehensive, as it can seldom be altogether complied with. But the following relationships should subsist:—

1st. Corresponding lines, whether representing constituent parts of the frame, or external forces, which meet at a point in the frame, form a closed polygon in the diagram.

2d. Corresponding lines which represent constituent parts of the frame, and form a closed polygon in it, meet in a point in the diagram.

3d. The lines representing all the external forces acting on the frame should form a closed polygon in the diagram.

4th. Lines—some of which represent external forces—which meet in a point in the diagram, have the corresponding lines contiguous in the frame; but these may form a partial boundary to an infinite area, as w_1, a, and P, Fig. 356.

5th. There should be only one line in the diagram of forces to represent any one force acting on the frame or in a part of it.

If the various diagrams to Fig. 356 be examined, it will be found that all except ii. fail in one or more of these relationships. Some of the diagrams may in themselves be equally clear and useful as ii.;* but were it for no other reason, we must prefer the reciprocal form, as admitting the use of a very simple method of lettering the figures, which will now be described.

NEW METHOD OF NAMING THE PARTS OF THE FIGURES.

This method, which I have contrived for lettering or naming the parts of the truss and its reciprocal diagram of forces, is founded upon that reciprocity. It will at once commend itself by its obvious simplicity, the assistance it affords in constructing the diagrams, and its doing away with any confusion arising from overlying or coinciding lines in the diagram.† This plan of lettering consists in assigning a particular letter to each enclosed area or space *in*, and also to each space (enclosed or not) *around* or bounding the truss, and attaching the same letter to the angle or point of concourse of lines which represents the area in the diagram of forces. Any linear part of the truss, or any line of action of an ex-

* *The Strains in Trusses computed by means of Diagrams, with Twenty Examples drawn to scale.* By Francis A. Ranken, M.A., C.E., 1872. In these examples Mr. Ranken adopts by preference diagrams which are not the reciprocal ones. In an appendix he gives two examples of reciprocal diagrams of forces.

† A further advantage will be found stated at page 57.

ternal force applied to it, is to be named from the two letters belonging to the two spaces it separates; and the corresponding line in the reciprocal diagram of forces, which represents the force acting in that part or line, will have its *extremities* defined by the same two letters.

EXAMPLE I.—Fig. 357 shows a complete truss and a complete diagram of forces so treated: here the letters A B C are assigned to the spaces in, and D E F to those around the truss i.; and also attached to the corresponding points of concourse in the diagram ii. The name of each line in the truss is also here given written against it, but this is quite unnecessary when the reader is acquainted with the method, each part in the truss figure being simply named from the spaces it separates.

Taking any triangle in the truss, say A, the stresses in the parts forming its sides are represented in the diagram by the three corresponding lines radiating from point A. The external space D of the truss figure must be regarded as an infinite area, but five contiguous lines are given, forming a partial boundary to it, and in the diagram we accordingly find the five corresponding lines radiating from point D, and their lengths represent the magnitudes of the respective forces. The *space* E is also infinite, and has only three defining contiguous lines, and the forces acting in these are given by the corresponding lines radiating from the *point* E in ii.

EXAMPLE II.—Fig. 358. This is nearly similar to the foregoing, but from making the reactions of the supports vertical, it is apparently simpler. It will afford us examples of overlying lines; thus space E is here only defined by the two vertical lines E D and E F, and in the diagram these are found proceeding from the point E, but overlying; similarly the reaction or supporting force D E partially overlies the weight F G in the diagram, these two lines have no terminal point in common. By these examples of overlying lines in the diagram, it will be seen that no confusion can arise under this mode of lettering. *Space* F has three contiguous bounding lines in Fig. i.; and in Fig. ii. we find the corresponding lines F E, F A, and F G, radiating from *point* F. The sides of the *polygon* A B C G F A in Fig. ii. represent the forces acting in the lines of the frame Fig. i., which meet at the *joint*, surrounded by the polygons and spaces A B C G and F, and so on.

DIRECTIONS FOR THE CONSTRUCTION OF DIAGRAMS OF FORCES.

1*st.* Assign a letter to each enclosed area of the truss, also to each division of the surrounding space as separated by the lines of action of the external forces.

2*d.* The lines in the diagram are to be drawn parallel to the corresponding lines or parts of the truss figure.

3*d.* The forces acting in lines radiating from a point in the truss must in the *reciprocal* diagram form a closed polygon.

Note.—The particular order in which the lines representing the forces are taken to form the closed polygon must be such that, when combined with the other polygons of the diagram, the reciprocity will be maintained. If the polygon is complete in itself, it is of no consequence in what order the lines are taken; thus let Fig. 359 i. represent a point acted upon by four forces in equilibrium; then the quadrilateral figure constituting the diagram of forces may be drawn in a variety of ways, each of which will be a reciprocal. Some of these diagrams are shown by Figs. ii. iii. iv. and v. of 359, and in each the directions in which the forces act against the point in i. are shown by the arrows arranged so as to appear to follow one another round the polygon. We require to know the magnitudes of two of the four forces before we can draw the diagram definitely. When one only is given, say for instance the vertical one, the values of the others may vary greatly, as shown by the dotted line in ii.

4th. The sides of any uncrossed space, triangle, or other polygon, around or in the truss, must always be represented by lines radiating from a point in the *reciprocal* diagram, and that point is to be named by the letter assigned to the space or polygon.

In diagrams which are *not* reciprocal, a triangle in the truss will also generally have its sides represented by three lines radiating from a point, but sometimes the lines are found connected, as in Fig. 360, or 361, or even detached.

5th. An important guide and check in constructing any diagram of forces for a fabric supported by vertical reactions is to balance the horizontal and vertical elements* of the forces cut by any line such as *a-a′*, Fig. 362. The sum of the horizontal elements of the intersected parts, which act as *ties*, must be equal to the sum of the horizontal elements of the intersected *struts*. In the Figure 362 D A acts as a tie, and its horizontal element will of course be just equal to the horizontal stretch of D A in the diagram ii., and F B and B A, the other parts, cut by *a-a′*, act as struts, and the sum of their horizontal elements is equal to their united horizontal stretches in the diagram, and the amount of this combined stretch is seen by simple inspection to be equal to the horizontal stretch of A D.

Then, as to the vertical elements, we can readily calculate the excess that must be in course of transmission past the section *a-a′* towards one or other support. In the figure there is evidently a downward pressure or shearing force in *a-a′* equal to half F F′ = F D in the diagram; then of the parts intersected, F B is carrying load down to the support E, while A B and A D are counteracting the work; therefore

* The term "element" is substituted for "component" when the two components of a force are taken vertically and horizontally. This convenient distinction was first employed in *Treatise on Bracing*, 1851.

the vertical element of F B vertical element of A B + vertical element of A D + F D, and this can be verified by attending to the vertical rises of the lines in the diagram.

6th. An important principle which the new method of lettering makes very evident is this. If we draw a line joining *any* two points or letters in the diagram, that line will represent in direction and amount the resultant of the forces acting in the parts of the fabric, and in the lines of action of external forces lying between the *spaces* corresponding with the same points or letters; and frequently we may take the passage from the one space to the other in a variety of directions. Thus in Fig. 132 ii., Plate XI., the dotted line C Y in the diagram represents the resultant of the forces acting in C D and D Y in the truss i.; or we may take other sets of parts and lines of force separating the spaces C and Y; thus C Y = resultant of the forces in C X and X Y, or in C X, X Z, and Z Y, or in C B, B Z, and Z Y. Now, in the figure under review, this principle, taken in a reverse manner, gives us a satisfactory way of determining the position of the point C in the diagram. Produce the line C X in Fig. 132 i. until it cuts X Y produced in the point *a*; now, by the principle illustrated by Fig. 343, Plate I., it is evident that a line joining *b* and *a* will give the direction of the resultant of the forces acting in C D and D Y. We need only therefore draw Y C in the diagram parallel to *b a*, and its intersection with X C gives point C.

The positions of the other points, such as E, can be determined in a similar way, or if found otherwise, the correctness of the determination checked.

7th. All the external forces acting upon a truss or other erection, must, taken together, be represented in the reciprocal diagram by the sides of a closed polygon.

The external forces are conveniently divided into two sets, the one comprising the forces to be borne by the structure, namely, the loadings, weight of structure, pressure of wind, etc.; the other, the supporting forces, or the reactions of the supports: the former are supposed to be known from the nature of the structure and the circumstances in which it is placed, the latter must be determined before we can complete the diagram of forces, and the varieties of the problem may be classified as follows:—

A. When the loading all acts vertically, and is uniformly distributed, and when the reactions of the supports are vertical.—In this, the simplest case, the two reactions will be equal to one another, and the two lines representing them in the diagram will form one continuous line overlying an equal line, made up of the lines representing the various parts of the loading, as in Fig. 358 ii.

B. When the loading all acts vertically, but is not equally distributed, and when the reactions are vertical.—In this case the two reactions or supporting forces will not be equal to one another, but the lines representing them will, as above, form one continuous line, overlying the line representing the loadings. (Fig. 56, Plate VI.)

The effects of each separate portion of the loading upon the two supports will be inversely as the distances of the supports from its line of action, and the total amount of the reactions are obtained by the summation of these partial effects.

C. When the loading is distributed in any manner, and all of it acts vertically, and when the reactions are inclined.—In this case the reactions must have opposite inclinations and equal horizontal elements: their vertical elements must be determined as in case B; and the polygon of external forces will be a triangle, of which the vertical side will represent the loadings (see Fig. 111, Plate IX.) The inclinations of the reactions may result either from the nature of the supports, or from the structure tending to increase or reduce the distance between its bearings: this may be resisted simply by the friction of the bearings, or intentionally as in abutting or suspension structures.

D. When the first set of external forces are irregularly distributed, and some of their lines of action oblique.—We may proceed by ascertaining the point for each supported force, where its line of action intersects a line joining the two supports, and there resolving it vertically and horizontally (when the line joining the supports is horizontal); the vertical elements so obtained are to be treated like the vertical loadings in case B; this gives us the vertical elements of the reactions. The horizontal elements of the forces resolved at the line joining the supports are to be algebraically added together; the residual horizontal force, if any, is the force tending to move the whole structure to one side, and must be resisted by the supports, but the proportions in which they contribute to this work will be determined by the constructive arrangements; thus the horizontal force would be nearly all resisted by one support alone, if that were a stout piece of masonry to which one end of the structure was fixed down, while the other end rested on rollers, or was simply propped up by a tall column. But if each support were equally stable, and the structure rested firmly thereon at both extremities, we might assume that the distribution of the horizontal force would be more equable, but liable to some variations from the elasticity of the parts and other causes.

Though it may frequently be best, it is not always necessary to calculate the proportions in which an irregular loading is distributed to the supports. The diagrams sometimes determine this at once, as for instance in Figs. 19, 20, 30, 32, etc.; or the position of the point corresponding with the space between the two supporting forces may be approximated to by successive trials.

APPLICABILITY OF THE METHOD OF DIAGRAMS TO THE DIFFERENT CLASSES.

The stresses in the various parts of trusses belonging to Class I. are obtainable at once from a diagram of forces, and the examples given will be nearly altogether confined to that class.

For structures belonging to Class II., a diagram of forces will give the stresses produced by an equilibrating loading, and may further assist the designer in estimating the severity of the transverse and other stresses produced by other loadings or forces. For structures belonging to Class III., the assistance to be derived from a diagram of forces will vary greatly; in some of these, by making certain assumptions, the stresses may at once be arrived at; in others tedious calculations may be required, in the carrying out of which, however, much assistance may be derived from the use of diagrams of the stresses. But the objectionable characteristic of many of the members of this class is the uncertainty of the calculated results truly representing the actual stresses in the structure when erected, from the conditions being or becoming different from those assumed.

The graphic method, as contrasted with numerical computation, will be found generally more suitable for roofs than for bridges. The first class of roofs will be very fully treated; and as the consideration of roof-trusses is made to precede that of bridges, and many of the remarks and modes of treatment are common to both, this will still further increase the apparent disparity in the space allotted to roofs. But if the reader familiarize himself with the examples given under roofs, he will experience little difficulty in projecting the diagrams for other structures. For some forms of parallel girders, although examples may be given, the graphic method is not so satisfactory as arithmetical computation, which will be treated of at length in another Part of the "Economics of Construction."

EXAMPLES OF DIAGRAMS OF FORCES FOR STRUCTURES BELONGING TO CLASS I.

Roof Trusses.

FIGURES 3 i. and 3 ii., Plate III., show a pair of untied rafters under two arrangements of the external forces. The corresponding diagrams of forces are given immediately beneath each drawing of the structure.

Figs. 3 iii. to 3 ix.—This couple or triangular frame, the simplest possible kind of tied truss, is here shown under a great variety of conditions of the external forces. In iii. there is only one force, C F, to be supported, and this is met by the two reactions C B and F B of the supports: all these three external forces being vertical, the polygon representing them in the diagram of forces degenerates into overlying lines. In iv. the reactions are assumed to be not quite vertical, and this assumption will be frequently made in the succeeding examples, as it shows clearly the polygon formed by the external forces; but in practice care should be taken to give the reactions the most prejudicial directions likely to occur. From the friction of the bearings at the supports, the reactions intended to be vertical may become somewhat inclined inwards at one time and outwards at another, leading to augmentations in the stresses of some of the parts above those produced when the reactions are vertical. In v. and the succeeding figures account is taken of the parts of the loading that are brought directly to the supports. In vi. the reactions are supposed to be inclined slightly inwards. In vii. this inclination of the reactions is made so great, that the tie-bar A B is required to act as a strut. In viii. the reactions take outward inclinations, or so as to augment the tension in the tie-bar. In ix. some of the forces to be supported are made oblique, to represent the combined effects of the loading and of a side wind. It will be observed that the diagram to 3 ix. embraces all the others except viii., and is reduced to one or other of these by assuming simpler conditions. If the stress in A B be made nothing, the diagram is that of ii. If D F and E F are made vertical, it becomes vi. If all the five external forces be made vertical, it becomes v.; and if, further, C D and E F be reduced to zero, it becomes iii.

Figs. 4 and 5 show the simple triangular frame when the supports are at different levels.

This very full treatment of the simple triangular frame, under a variety of conditions, will render it unnecessary to give more than one or two illustrations of the trusses to be afterwards chosen as examples. There is another condition—namely, that of irregularly-distributed loading—which the simple triangle is not suited to illustrate. Diagrams showing the effects of this in augmenting the stresses in some of the parts over what they suffer when the whole loading is on will be found under Figs. 19, 32 iv., etc. Similar results of course also arise in such structures from the lateral action of the wind.

No. 7, Fig. i. gives the truss under vertical loadings and slightly inclined supporting forces. Fig. ii. shows it under the combined effects of dead-weight and side-wind pressure.

Fig. 9.—This truss is here assumed to be subjected to vertical loadings, placed both on the rafters and the tie-beam. The diagram could be drawn otherwise, by regarding the lower loading as brought to bear on the tie as downward pressures instead of as downward pulls. Both modes are illustrated by Figs. 61 i. and ii.

No. 10.—Fig i. shows the truss under the dead-weight

alone. In Figs. ii. and iii. the effects of a side wind are superadded. In ii. the leeward foot of the truss is supposed to be fixed, and the windward one movable, as when placed on rollers. In iii. this arrangement is reversed. Attention should be directed to the greatly augmented stresses under the conditions shown by Fig. iii.

No. 11.—This truss does not require the brace A B when placed under a symmetrical loading, but under an irregular loading, or when a side wind has to be resisted, it comes into action. The diagram shows A B acting as a strut when F G is rendered oblique by a side wind. Of course when the wind blows from the opposite quarter, A B must do duty as a tie. The effects upon A B of a difference in the loadings on the two sides of the truss are illustrated by Fig. 19.

No. 19.—Diagram ii. is for the case of F G being greater than H I. Diagram iii. shows the stresses when F G is less than H I. The signs + and - placed against A C indicate strut and tie action respectively.

Many of the diagrams will be passed over without comment; but as they are generally given to illustrate some point of interest or importance, they may be thought deserving of the studious attention of the reader.

No. 32.—This design, when deprived of a tie-bar, is only suited for an abutting principle, or when the length of span is fixed without its own aid. Diagram ii. is for the case of the right-hand side of the truss being fully loaded, the left-hand side with half the load removed, and a side wind acting from right to left. Fig. iii. gives the diagram for the full uniform load, and shows how nearly the arched rafter is equilibrated under such. Fig. iv. is the diagram of forces when the load is supposed to be altogether removed from the left-hand half of the structure. Fig. v. gives the effects of the wind taken alone, and of about twice the previous intensity, and according to the ordinary theory that the pressure at right angles to an inclined surface varies as the square of the sine of the angle it makes with the plane of the horizon. Fig. vi. is the same, on the assumption that the pressure varies simply as the sine of that angle, an assumption that is perhaps practically more correct; but the subject is one that has been much neglected, or unsatisfactorily investigated by experimentalists.

No. 40.—This design is particularly well suited for iron work. When the supporting forces are in the directions of G to C and C to L in the diagram, the tie C F is free from stress, and the case becomes that of No. 46.

No. 61.—This is 46 tied, and supplied with suspension rods going down to the main tie, which may be supposed to support a floor. In Fig. ii. are shown the changes in the lettering and form of the diagram caused by treating the loadings supported at points a and a', as though imposed by downward thrusts instead of downward pulls.

No. 70.—The leeward foot is here supposed to be placed on rollers, which arrangement, as we found in considering No. 10, gives the greatest stresses.

Nos. 88, 89, and 90.—The first of these is somewhat out of place here, but the three are brought together in order that a comparison may be readily drawn as to the effects produced by changes in the outline of the main tie. In the modification 88, it will be observed that the lowest part of the rafter A M is subjected to a much greater stress than the highest part, E D; and as the number of triangles is increased, the proportion tends to become as 2 : 1. Now, in practice, it is found convenient to make the sectional area of the rafter uniform throughout, and of course it must be in proportion to the stress in the lowest part here. The sectional area of the rafter is usually calculated with an allowance of metal per ton of stress double that given to the main tie. There is consequently a considerable amount of material uselessly applied; and this loss is still more important when, as is very usual, an upward curvature is given to the tie-bar, as in No. 89. Now, I would suggest that the tie-bar should be curved, as in No. 90, when doing so would lead to no inconvenience. Such a form renders the stresses in the rafters more uniform throughout, and the sectional area required is very greatly reduced. The material required for the other parts would also be less.

This suggestion has also been embodied in the modifications Nos. 50, 59, 69, 84, 98, 104, and 120.

Nos. 106 and 116.—The figure given on Plate X. for these shows one half made after the one pattern, and the other half after the other: this is merely done to save space; the two diagrams of forces are given beneath the figure of the truss, combined also, but so as need not lead to any confusion.

The figures and diagrams for several other designs are similarly shown in combination.

Nos. 127 and 128.—In the drawings of these trusses the lines or arrows to represent the portions of the loading, etc., have been omitted; they are not necessary, since the magnitudes and directions of these external forces appear in the diagrams, and they somewhat disfigure the designs.

It appears unnecessary to add examples with more numerous regular subdivisions of the span: little difficulty should be experienced in applying the method to such, if the examples given be properly studied.

No. 131.—This is one of the trusses of Class I. characterised by the irregularity of size of the subdivisions of the span, and by the number of these being greater than required for the support of the roof platform. The diagram for this truss has been omitted, but the modes of delineating it will be gathered from the account of the next example.

Under No. 88 it has been pointed out that the lower end of the rafter in the roof with outlines such as this is more severely stressed than the upper end. In No. 131 some compensation is provided in the more perfect stiffening of the rafter towards its foot, caused by the shorter intervals between the supported points.

No. 132.—From the minuteness of the triangles at the extremities of this design, we do not get a good starting basis in the usual manner exemplified by Nos. 105 and 118. There are several methods of overcoming the difficulty; we may make a temporary assumption that the truss finishes with a larger triangle substituted for the smaller terminal ones; and after properly transferring the loading on the part occupied by it, proceed to draw the diagram. The lines representing the stresses in the part of the small triangles may thereafter be inserted, the letter assigned to the temporary triangle being rubbed out. The terminal point W, 132 ii., is easily obtained by drawing lines parallel to the tangents of the arch and tie-bar at the foot of the truss.

Another method of proceeding is indicated in "Direction" 6th, page 57; by this we begin at any part; the mode of ascertaining the resultant *b a* of 132 i., for a weight at *b* there described, is equally applicable to any loading confined to one side of *b*, or any other point. In 132 i. and ii. the resultant C Z is parallel to *b c;* resultant B X parallel to *d e,* and so on. To obtain by this method the stresses for the case of the whole load being on, two diagrams may be drawn—the one representing the stresses produced by the load lying to one side of any point *b*, and the other diagram the effects of the load on the other side of the same point, the load or force at *b* being either divided between the two parts of the loadings, or included wholly in one of them; the two stresses so obtained for each part are to be algebraically added. The effects of any other distribution of the external forces or loading may be arrived at in a similar manner by combining the stresses given by its subdivisions.

A third method is to ascertain the position of a convenient

starting-point, such as A or B, by calculating its distance, A X or B Z, from the known points X or Z, by equating the moments measured round any suitable point; otherwise we may use the result thus obtained as a check upon other methods.

132 iii. and iv. illustrate the case of a side wind acting in combination with a uniform loading. It is obvious from iv. that when the whole truss is not loaded very unequally, each brace acts as a tie; and before a strut action can be induced, this tie-action must be neutralised—that is, each brace can act virtually as a strut to the extent of its previous tension, without suffering compression in doing the work of distributing any superadded irregular loading. This is a source of economy in wrought-iron work, since it is cheaper in such to provide ties than struts, especially when the latter are long in proportion to the amount of stress to be conveyed.

No. 137 is the roof-principal of the Drill Hall recently erected at Forrest Road, Edinburgh; I supplied the architect with this design in 1871, and assisted him in carrying it out. The span for calculation is 97½ feet. The diagram given in Plate XI. is for a uniform loading, and drawn by assuming that the smaller triangles towards the extremity are replaced by triangle *s*, according to the method described under No. 132.

EXAMPLES OF DIAGRAMS OF FORCES FOR STRUCTURES BELONGING TO CLASS I.

Bridges.

It has been already noted that a design for a framed bridge may be used in two positions; if the one be considered as the erect position, the other is called the inverted one. Now one diagram of forces is sufficient to give the stresses in the parts of the truss whether in its original or inverted state. If we turn the truss simply over, so that what was the top becomes the bottom, the diagram, if the same lettering be retained, requires to be turned so that what was the right-hand side becomes the left, as in Figs. 201 i. to iv., Plate XII. But if, in addition to turning the truss upside down, we also turn it over right to left, as shown by Figs. 205 i. and ii., the one unchanged diagram 205 iii. will be suitable in all respects for both arrangements.

Although the design be inverted, the directions in which the external forces act are of course not reversed, except relatively to the design; but to suit the resulting changes of struts into ties, and ties into struts in the truss, the external forces that appear to be applied as thrusts in the one, may, if thought desirable, be changed in appearance to pulls in the other drawing, as shown in Nos. 205 and 219.

THE DIRECTION OF THE REACTION OF THE SUPPORT.

In all cases of girder action in which the truss rests upon horizontal surfaces, it is usual to assume that the reaction is vertical. Now, although it always approximates to this, yet it is evident that when there are any influences at work tending to cause the girder to slide on its bearings, the resistance which the friction offers to this may produce a considerable deviation in the direction of the supporting force when no friction rollers are used. In the case of a simple parallel girder with its bearings at the level of the lower boom, this, when acting under its own weight alone, may be supposed, from the circumstances of its erection, to rest vertically on the supports at ordinary temperatures; but when the movable loading comes upon the girder, the length of its bottom boom is increased, and this will usually be only in part taken up by the curvature: the bottom part of the girder therefore becomes lengthened in span if it still act as a simple girder, but this lengthening is resisted by the friction on the bearings, and may be altogether or in part prevented, and the reaction in consequence will be inclined inwards, thereby reducing the tension in the lower boom below what it would be were the supporting force truly vertical.

When a girder that strongly evinces this tendency to lengthen the distance between its bearings is prevented doing so by pinning its ends, the reactions may take a considerable inclination inwards when the extra load comes on, especially if the structure be of iron, and the temperature be much higher than when the pinning was effected. In other circumstances, when the load is off, a fall of temperature may cause the reactions to incline outwards: of course when the ends are restrained by the friction alone, these inclinations cannot exceed the angle of repose, and would probably indeed be confined to a much smaller deviation in the presence of the tremor caused by a passing load.

In the case of the triangular girder, Fig. 205 i., the length between the bearings tends to become reduced from the compression of the upper member. This will cause the reaction of the supports to slope more or less as shown ; the stress in the top, as at A F, will be thereby reduced, while the tension in the tie A D remains unaltered. When the same truss is used in the inverted form shown by Fig. 205 ii. the action of the load will tend to increase the span, so that the friction of the bearings will cause the reactions to assume the inward slopes, relieving now the tie A F of some of its tension, but without affecting the upper member A D.

It is fortunate that in all these cases the effect of the inclination given to the reactions from this cause is in favour of the strength, of parts at least, of the truss. It must be borne in mind, however, that a change of temperature may produce an opposite effect.

No. 202, A, B, and C.—Three arrangements of the forces, and corresponding diagrams of stresses, are here given to represent the supporting of the central weight: in A, through simple bracket-action ; in B, through simple abutting action ; and in C, by the structure acting as a tied truss.

In actual structures the mode of action is often a complicated one compounded of these—

No. 211.—The unsatisfactory character of this description of truss is shown by the excessive stresses brought upon C I, A B, etc.

No. 220 A is the case of the loading being placed on the longer member. Here, for simplicity, the parts of the loading which would be imposed directly upon the supports are omitted.

220 A i. is the diagram of forces when all the other points are fully loaded.

220 A ii. is the diagram when two-thirds of the loading is removed from point K L.

220 A iii. is the diagram when two-thirds of the loading is removed from each of the points K L and J K.

No. 220 B gives the case of the loading being imposed at the level of the shorter member.

220 B i. is the diagram when all the length is fully loaded.

220 B ii. the diagram when three-quarters of the length is fully loaded, two-thirds of the loading being removed from the other quarter.

220 B iii. the diagram when half the length is fully loaded, the other half being loaded to one-third.

220 B iv. the diagram when one-quarter of the length is fully loaded, two-thirds of the loading being removed from the remaining part.

No. 231.—This, compared with No. 230, has double the depth of structure, but the saving of material when designed with similar angles is less than five per cent., No. 230 is therefore the better design of the two, as its depth could be increased without inconvenience.

No. 261 (and a modification of No. 262).—The diagram of 261 may be constructed by first composing the partial one 261 iv. of the part of the frame shown by 261 iii.

No. 263½. (This is just No. 266 with eight instead of nine subdivisions of the span.)

This design is only interesting when the supporting forces are much inclined, or, in other words, when the structure becomes abutting. But when abutting, the degree of inclination of the reactions is a subject practically of much doubt, the structure belonging then to Class III., and exhibiting the defects of that class in a high degree when erected in the usual way. But for any definite assumption as to the directions of the reactions of the supports, the structure is to be dealt with as belonging to Class I.

263½ ii. is the diagram of forces when the supporting forces are vertical.

263½ iii. is the diagram when the reactions are horizontal—that is, when it is subjected alone to two horizontal forces acting in the line R *r*, and tending to bring the feet of the structure nearer together.

263½ iv.—The diagram of stresses, when loaded uniformly, and a joint in I R is opened to allow of a more accurate adjustment of the structure to the span, as suggested by me some years ago.

No. 267.—As in 261, so here, a little calculation is required, or a subordinate diagram of forces must be drawn for any irregularly loaded sub-frame, as H M L here.

APPLICATION OF THE GRAPHIC METHOD TO STRUCTURES OF THE SECOND CLASS.

SOME EXAMPLES OF THIS CLASS OF IMPERFECT STRUCTURES ARE PRACTICALLY USEFUL FOR ROOFS.

ALTHOUGH the designs of this class may not, when considered in the usual theoretical manner, be capable of acting as rigid or stable trusses under a non-equilibrating loading, yet, through being connected, by means of the planking and other parts running at right angles to the plane of the truss, with partitions and gable walls,* and by having their joints made very stiff against turning, and receiving a very liberal allowance of extra material to compensate for induced cross stresses, they may become practically very satisfactory as regards stability. And when the increased outlay thus incurred is repaid by the value of the uninterrupted openings given by such forms as Nos. 139, 145, 153, 159, etc., they become important, and deserving of scientific attention. As, however, the graphic method can only be of subordinate use in dealing with such structures, it would not be suitable to enter very fully into their consideration here.

The stiffness of the joints adds to the power of resisting distortion by constraining the parts to assume much more complicated flexures. I shall, however, in what follows,

* In the cases of Nos. 138, 139, 149, and other designs, the boarded platform on the top may be made to act as an efficient horizontal bracing against distortion. In iron structures it might in many cases be advisable to supply proper horizontal or vertical bracings running at right angles to the planes of the trusses.

neglect the advantage to be so gained, and take note only of the simpler flexures that would be produced if the joints at the ends of the pieces meeting the principal members of the frame were hinged, and regard the stability under a non-equilibrating loading as altogether resulting from the transverse stiffness of such principal parts.

DIAGRAMS OF FORCES FOR STRUCTURES OF CLASS II.

I shall now describe a mode of drawing a diagram to give the *longitudinal* stresses in an imperfect truss subjected to a distorting load, and therefore dependent for its stability upon the transverse resistances of some of the parts. This consists in supplying imaginary forces acting on the frame, which shall represent the reactions against bending, and then proceeding with the diagram as though these were real forces, and the parts freed from all transverse stress. The severity of the transverse stress in any member must be dealt with subsequently by considering the part as subjected to forces equal and opposite to those imaginary forces in addition to the longitudinal stresses given by the diagram.

This mode of treatment will be most clearly explained by applying it to a suitable example, such as shown by Fig. 339 i., Plate XVI.

The black arrows show the external forces having a real existence, the dotted arrows the imaginary forces to represent the reactions of the tie-beam against being bent. Let G H be taken as the unit of force, and let F G = 2; now half of F G, acting along with G H, would constitute an equilibrating loading, so that the distorting force is the half of F G, and the duty to be performed indirectly by the transverse stiffness of the tie-beam is to convey one-third of a unit from F G to the right-hand support. The necessary values to be assigned to the imaginary forces are therefore such as written against them. The diagram of the full longitudinal stresses next to be drawn is shown by Fig. ii.; and the transverse forces acting on the tie-beam are represented in Fig. iii.

The results are somewhat remarkable: the longitudinal stress in A F is somewhat, and the stress in B H is very considerably, greater than would be the case were the structure efficiently braced, arising from the *lifting* effect of the tie-beam acted upon by the tension in A B.

In more complicated cases, or generally, it will be most convenient to treat such structures in two operations—1*st*, by drawing a diagram of the longitudinal stresses produced by all that part of the loading that would produce equilibrium; and 2*d*, by a second diagram to show the separate effects of the disturbing part of the loading: the stress for each part being finally arrived at by the algebraic addition of the separate results. As there can be no difficulty in drawing the first diagram, I shall in the following examples consider the effects of the distorting force alone.

EXAMPLE 2.—Truss No. 139 (Plate XVI.)

Fig. 139 i. shows in an exaggerated manner the nature of the distortion under an irregular loading.

It may be assumed that the sides of the opening B will always form a parallelogram, and therefore that the stresses in A B and B C will be equal.

The position of the point D or F horizontally in the diagram iii. depends upon the horizontal elements of the reactions of the supports. The only effect of increasing or reducing the relative width of the opening B is to increase or reduce the values of G H and I J compared with F E or E D; thus

$$\text{G H} = \text{F E}\,\frac{\text{width of B.}}{\text{whole span.}}$$

The stress in A B

$= \frac{1}{2}$ vertical element of H I + horizontal element of

$$\text{H I}\,\frac{\text{Length of A B in ii.}}{\text{2 length of A F in ii.}}$$

The vertical element of the stress in H A = the difference instead of the sum of the above two quantities.

When H A is inclined at an angle of 45°, and H I is at right angles to H A, the stress in H A = O.

As in timber structures the tie-beam usually supplied is stout, and well calculated to resist the flexures shown by Fig. iv., the comparative efficiency of this truss is easily accounted for.

It is of some practical importance to note that the stress in C I is greater here than when a diagonal brace is inserted in the quadrilateral opening.

EXAMPLE 3.—Truss No. 141 (Plate XVI).

Fig. i. shows the nature of the distortion to be resisted. Fig. ii. is the truss with the distorting and imaginary forces applied to it. Fig. iii. the diagram of longitudinal stresses; and iv. the forces causing flexure in the rafter.

The central bending force on each rafter will be equal to half the component F *a* of the force F G resolved in the direction of, and at right angles to, the rafter.

When the supporting force C K is vertical, the points C and A will coalesce in the diagram. If C and B coalesce in the diagram, there will be no stress in the tie-beam, and the supporting forces will act in the directions of the rafters.

EXAMPLE 4.—Truss No. 145 (Plate XVI.)

When the resistance to bending of the rafters may be neglected, the case is that of Example 2, Truss No. 139.

When the resistance to bending of the tie-beam is neglected, the case is the same as Example 3, Truss No. 141.

One portion, say X of K L, will be supported on the first mode of action, and the other, say Y, on the second; the relative amounts of X and Y will depend upon the relative stiffnesses of the tie-beam and the rafters. In wooden structures, as usually made, X will be several times the value of

Y; but in iron work, where a round bar of iron is used for the main tie, X will be of trifling value compared with Y; and as the rafters are ill suited to withstand the simple transverse flexure, the design is a bad one for a fabric of iron, but a passably good one when wood is the material employed.

When the width of opening C is one-third of the span, Q E or H I = $\frac{1}{3}$ E F.

Note.—A transverse stress applied to a long strut is very prejudicial to its strength as a strut. A transverse stress applied to a tie-beam is much less objectionable.

ANOTHER METHOD OF APPLYING DIAGRAMS OF STRESSES TO STRUCTURES OF CLASS II.

In dealing with any truss of Class II., which could be converted into Class I. by the addition of one brace, another mode of proceeding may sometimes be adopted, and it may be used for cases in which there is a complicated departure from equilibration in the arrangement of the loading or other external forces.

Let the brace requisite to convert the truss into Class I. be supposed to be temporarily added, and draw out a diagram of forces for the structure as though it were of Class I. Next, instead of the real external forces, insert the imaginary ones to represent the reactions against cross-bending, and draw out a second diagram, and let the scale of this be such that the stress in the temporary brace will be the same in magnitude as in the first diagram; it will, however, be of opposite character. Then, to arrive at the final stresses, add the results in the two diagrams algebraically. The temporary piece will come out of course free from stress, and so may be supposed to be withdrawn.

This method might perhaps be extended to cases in which more than one brace would be required to render the structure perfect.

EXAMPLE 5.—Truss 279 (Plate XVI.)

The following example of this second method is of a very simple character, having only one distorting force, E F, Fig. 279 i. Fig. ii. is the first diagram of forces obtained by treating Fig. i. as converted into Class I. by the temporary addition of the brace B B. Fig. iii. shows the frame under the imaginary forces; and iv. is the second diagram of forces corresponding with it, drawn to such a scale that B B′ in it will be equal in magnitude to B B′ in Fig. ii. On adding the stresses for the same parts in diagrams ii. and iv., it will be found that the final stresses in A D and C D are equal; for (attending to the vertical elements alone since the parts are equally inclined) C D $= \frac{1}{3} + \frac{1}{6} = \frac{1}{2}$, and A D $= \frac{2}{3} - \frac{1}{6} = \frac{1}{2}$, E F being accounted $= 1$.

STRUCTURES BELONGING TO CLASS III.

OR THOSE HAVING A REDUNDANCE OF PARTS, AND THEREBY OFFERING MORE THAN ONE CHANNEL FOR THE CONVEYANCE OF THE FORCES.

ABUTMENTS and TIE-BARS.—*These play nearly the same part in Structures of Class I. Their differences in Class III.*

IN the structures of the first class it is not necessary to take into account the elongation of the tie-bar in tied trusses, or any moderate change of span in the case of abutting ones; and therefore the horizontal elements of the reactions of the abutments and the tension of a straight tie-bar at the level of the supports may, for such structures, be regarded as exactly interchangeable means of withstanding the horizontal thrusts.

If to any such structures as 105, 266, etc., of Class I., which are complete girders, but tend to spread somewhat, we substitute stable and perfectly fitted abutments for simple supports, it is much the same theoretically as though we added very stout horizontal tie-bars, and either addition removes the design into Class III.

In Class III., however, an important distinction must be drawn between these arrangements in practice. When a tie-bar is employed, its elastic elongation under a known stress may be exactly calculated, and the stresses in the structure may therefore, when the fitting is carefully attended to, be estimated with some degree of confidence. But when abutments are used, we can place no trust in calculations based on the misleading theoretical assumption that the magnitude of the span may be regarded as known and permanent.

PERFECT FITTING IN SOME STRUCTURES BELONGING TO CLASS III. IS AN IMPOSSIBILITY UNDER DIFFERENT DEGREES OF LOADING.

Let us take No. 167 as a first example of Class III.; this is No. 15 of Class I., with a redundant tie-bar added. In iron work it is almost necessary to treat the diagonals as capable of acting both as struts and ties.

The distribution of the stresses in this Class is determined by the changes of length of the parts from elasticity when acting under the stresses, and as no one part can be lengthened or shortened without inducing stresses in other parts, the stresses may result from the combined effect of misfitting and the action of the loading. For the truss under consideration, when of the proportions shown in Plate XV., the following statement of results will give some idea of the interaction of the parts.

If the structure be put together so that, when unloaded, there shall be no initial stress caused by misfitting of the diagonals, etc., then, on adding equal weights, F G and G H, different effects will be produced on the diagonal braces, according to the allowance of material in the parts F A, B G, and D H, to carry each unit of stress, compared with the allowance in C E. *First,* if the first-named parts be made excessively stout, while the allowance of metal in the tie C E is small, we may consider that of the parts forming the boundary of the frame, C E is the only one that undergoes a change of length when the truss is loaded; the result of this will evidently be that the diagonals must become lengthened, and consequently both the braces A C and C D subjected to tension. *Second,* If, as is very usual in iron roofs, the stress per sectional inch on the rafters and top piece is only half that upon the tie-bar, the diagonal pieces will be almost free from stress whether the truss be loaded or unloaded. *Third*, If the allowance of metal to the strut and tie pieces be at the same rate per ton of stress, the diagonal braces will be subjected to compression when a uniform loading is added.

Again, if there be an excess of loading at one side, say at F G, about seven twenty-fourths of this excess must be conveyed to the support H E by means of A B acting as a strut, and A C as a tie; the arithmetical sum of the vertical elements of the stresses in these must be $= \frac{7}{24}$ (F G – G H), but in what proportion this is distributed between them depends upon the allowance of metal in the rafters and main tie. If the rafters have twice the allowance of the tie-bar, the stresses in the diagonal braces will be nearly equal; if the same allowance, the stress in A B will be greater than in A C. But we cannot assume the existence of perfect fitting, more particularly in large structures not previously put together when lying flat on the ground, and while properly supported at every point. The tie-bar C E, for instance, may be put on too loosely, or too tightly, and thus cause the braces to be subjected to tensions or compressions under a symmetrical loading from that cause alone; indeed, if put on too tightly, all the parts are subjected to stresses independent of any load.

OBJECTIONS TO CLASS III.

The fact of misfitting must frequently be so difficult to detect, and its influence upon the stresses may be so important, that the judicious engineer may well entertain an aversion to employing many of the designs belonging to this class, when those of the former classes will suffice. Some of the examples of Class III., are, however, much less open to objection than others; and whether they should be condemned must depend upon the practical difficulties and sources of error connected with them. There might be no good grounds for entertaining apprehension of evil in employing such a form as No. 174, which is indeed only a modification of a girder with two series of triangular bracings; and the evil influence of misfitting almost vanishes in such structures as Nos. 323, 324, etc.; but the danger becomes much more serious when a structure such as No. 132 is to be used as a large *abutting* roof principal, or No. 266 as an abutting bridge. There is, first, the difficulty of fitting it to the span between the abutments; second, the uncertainty of that distance between the abutments remaining of the estimated extent under the various influences of the thrusts, the settlements of the masonry and foundations, the action of a watercourse or drain passing the footings, the pressure of earth behind the abutments, etc. etc.; and, third, the expansion or contraction of the superstructure from change of temperature.

DIRECT USE OF DIAGRAMS OF FORCES IN CLASS III., NO. 195.

We might go through the form of drawing diagrams of the stresses for such structures as 195, etc., by assuming that the diagonals can only act as struts or as ties, and further, that only one of each pair can act at one time. The first assumption would generally be materially incorrect in iron work, and the second is only compatible with perfect fitting of all the parts, and not always even then, as we have seen in the case of truss No. 167. The error, from the first assumption, would, however, be on the side of safety. But making these assumptions is really just reducing the structure to Class I. for each operation, and there should now be no need to give further illustrations of that class.

No. 195 is the truss or principal used for the great roof at Birmingham; the span of the largest truss is 211 feet. When occupied in taking out the stresses in this for the

Messrs. Fox and Henderson (Sept. 1851), I drew Mr. C. H. Wild's attention to the fact that, with the circular curves adopted, and with the structure uniformly loaded, the stress on the main tie was greater at the centre than near the extremities. (This may be seen in Fig. 137 ii., where B X is evidently greater than S X or W X.) In reply, he showed me the following interesting relation that subsists between structures with arched and straight ties

If we draw beneath the diagram 340 i. (Plate XV.) showing the truss with arched tie, another, 340 ii., having a straight tie-bar and the same span, and also the same height of framing at corresponding points in the span, as for instance, $m' n' = m n$; then the horizontal elements of the stresses at all corresponding points in the two trusses under similar loadings will be equal. If the arch ii. were a parabola, both fabrics would be equilibrated by uniform loadings, and either structure under such a loading could do without bracing, perpendicular ties only being necessary.

Any curve might be chosen for one of the longitudinal members, the other being drawn so as to fulfil the relation with regard to depth of structure. Figs. iv. to viii. show some curious varieties of trusses thus drawn, the depths being regulated by the parabola iii. From a property of the parabola it follows that, if the arch or tie be made parabolic, as in iv., both must be so. It must not, however, be supposed that equilibrated trusses are here put forward as the best. The effect of side pressure from the wind has much to do with the economic question, as may be seen by studying the diagram 132 iv.

THE STRESSES IN AN ABUTTING STRUCTURE OF CLASS III.

Fig. 166 i., Plate XV., represents the simplest form suited to illustrate some of the peculiarities of this case. It is easily shown that under the condition that the span of the truss is to remain constant, we cannot render the sectional areas proportionate to the stresses; for if we assume the sections in proportion to the stresses, each part (in whatever proportion we take the sectional areas, for this has no effect upon the question) must undergo a change of length proportionate to its length; and as all the parts are under compression, there will be no change of shape produced; the truss will remain of exactly similar shape, but of a reduced size; and therefore, in order to again extend the span to the original length, we must (if all the load be already supposed to be on) apply extending horizontal forces at the level of the supports; but this will necessarily reduce the compression in A C, and increase that in A D, and therefore render the stress per sectional inch greater in A D than in A C.

There are a great many modes in which this excess of stress in the upper parts may be exhibited. We might regard the line joining the abutments as a neutral axis; or we might calculate the compressions—1st, on the supposition that A B does not become shortened, in which case the stress per inch on any oblique piece, D A or A C, produced by a given descent of point, *w*, will vary as $\sin. \varphi \div \sec. \varphi = \sin \varphi \times \cos \varphi$, φ being the angle made with a horizontal line. This increases up to a value of $\varphi = 45°$, at which angle it becomes a maximum;* so that, if the inclinations be both less than 45°, the uppermost piece will suffer the greater stress per sectional inch. When the span and A B are of constant lengths, the stresses per sectional inch on A D and A C will be as 5 : 4, in the form as drawn here. Then, 2d, it is evident that the effect of the shortening of A B will be to cause a still greater proportion of the stress to come upon the upper strut.

* I showed, in 1851, that the most economical angle for the braces in lattice girders was 45°. It appears from the above that 45° is also the angle at which the greatest stiffness is attained. More will be said on this point in another part.

But the most satisfactory way of dealing with this and some more complicated structures of this kind, is to assume in succession two different directions for the reaction of the abutment, and calculate for each the change caused in the length of the span; the reaction or supporting force that will cause no change of length in the span is then easily ascertained by taking for its components such proportions of the two assumed reactions, that their effects in altering the length of the span will neutralise one the other. If both the assumed reactions produce the same kind of change on the span, it is only requisite to treat one of them, and the corresponding loading or forces to be resisted, as acting in the opposite directions—that is, it must be taken negatively. In practice it will generally be convenient to take one of the reactions as vertical or horizontal, while the other may approximate to the correct inclination; or we may make one vertical and the other horizontal.

The following table gives the results for four different assumed directions for the reactions of the abutments; the sectional areas of all the parts are made the same. V is the vertical, and H the horizontal, element of the reaction:

Direction of the Reaction D C.	V.	H.	Change produced on the Span.
Vertical . .	0.5 w	0.	+ 6.1
D A	0.5 w	0.5 w	+ 1.4
A C	0.5 w	1.0 w	- 3.3
Horizontal . .	0.	0.5 w	- 4.7

Adopting the first and fourth trial reactions, then, in order that the span may not change when $V = 0.5\ w$, we must evidently take—

$$H = \frac{6.1}{4.7} \text{ times } V = 0.649\ w,$$

and the height of point x above a will be $= \frac{5000}{649} = 770.4$

$$x\,w = 229.6$$
$$\text{and } x\,b = 270.4$$

The diagram of forces for this reaction is shown by Fig. 166 ii.; in it D $a = 0.5\,w$ and C $a = 0.65\,w$.

If we suppose the first two trial reactions given in the table to be those adopted, we see that each of these causes elongation of the span; therefore, taking the first negatively, or acting as a downward pull and the loading acting upwards, the magnitude of this required to neutralise the 1.4 of extension caused by the second reaction will be $= -\frac{1.4}{6.1}\,\frac{1}{2}\,w$; combining this with the second trial reaction, having $V = \frac{1}{2}\,w$, and $H = \frac{1}{2}\,w$, we get a resultant with $V = \frac{2.35}{6.1}\,w = 0.77\,\frac{w}{2}$ and $H = \frac{1}{2}\,w$, which will have the direction of the reaction sought for. We must therefore take in that direction a length that will have $V = 0.5\,w$, and therefore $H = \frac{1}{2}\,w \div 0.77 = 0.65\,w$.

Or, again, taking the second and third trial reactions, it is evident that neutralization of effect on the span will be produced when the components of the reaction of the abutment, taken in the directions of D A and A C, are in the ratio of 33 : 14.

The dotted line joining x, with the two supports, is the representative line of pressures; and from a property of such a line the stresses in A D and A C will be inversely as the lengths of the perpendiculars xy and xz, let fall upon them from any one point in the line. Calculated thus, or measured from the diagram, the stresses per sectional inch of A D and A C come out very nearly in the ratio of 3 : 2.

If a certain change of the span were assumed, we could readily ascertain the corresponding position of the line of pressures. If the span be extended, the stress per sectional inch on the upper member, D A, will be still further augmented, compared with that of the lower one, A C.

No. 327.—This, or No. 266 of Class I., used as an abutting structure, was given in my Treatise on Bracing in 1851; and therein, without entering into any particular calculations for it or No. 328, I indicated, by means of a line of pressures in each figure, the peculiarities of the distribution of the stresses. I at the same time endeavoured to impress the fact that little confidence could be placed upon the line of pressures assuming and maintaining its theoretical position, and stated the practical precautions that were in consequence advisable.

But in the years 1870 and 1871 I suggested a mode of erecting such a bridge as No. 327, that would in a great measure do away with the doubtful character previously attached to it. The line of pressures would thereby be compelled to take up a definite position, and that position the most economical one. This contrivance, which has already been noticed under No. 263½ iv., may receive a fuller description in a subsequent part of the work.

As in the case of No. 166, the sectional areas cannot be made proportionate to the stresses in No. 327 as ordinarily treated, unless a *contraction* of the span take place. When the span remains the same, or becomes elongated, the stress per sectional inch must necessarily, in all designs of proportions likely to be employed, be greater in the upper than in the lower longitudinal member at the mid-span. It is the more necessary to call attention to this, as very erroneous views have been published on the subject.

PLATE I

341

342

343

344

345

346

347

348

349

350

351

352

PLATE II.

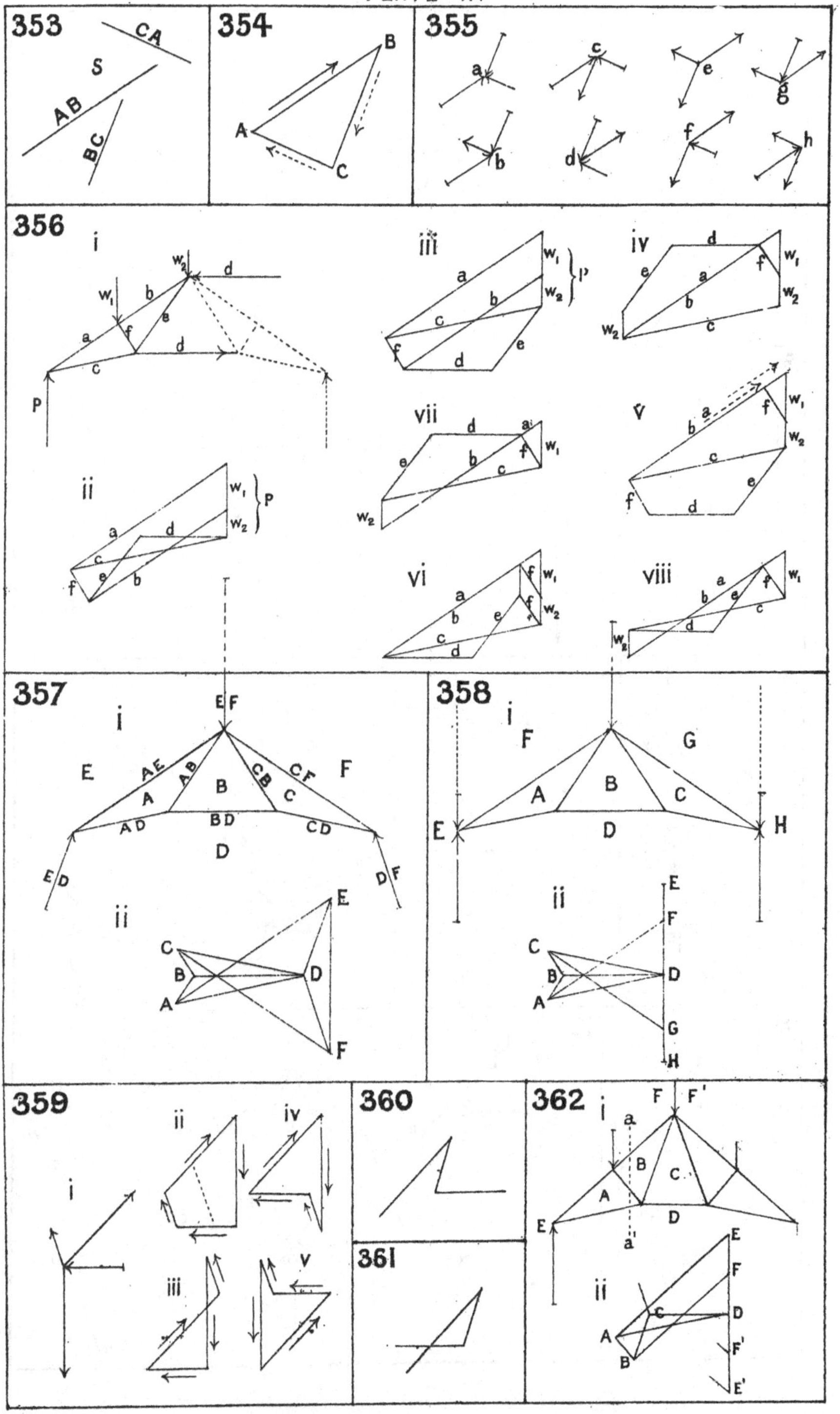

353
CA
S
AB
BC
354
A
B
C
355
a
b
c
d
e
f
g
h
356
i
ii
iii
iv
v
vi
vii
viii
357
i
ii
EF
E
F
AE
AB
CB
CF
A
B
C
AD
BD
CD
D
ED
DF
358
i
ii
F
G
A
B
C
E
D
H
359
i
ii
iii
iv
v
360
361
362
i
ii
F
F'
a
a'
A
B
C
D
E
E'

PLATE III.

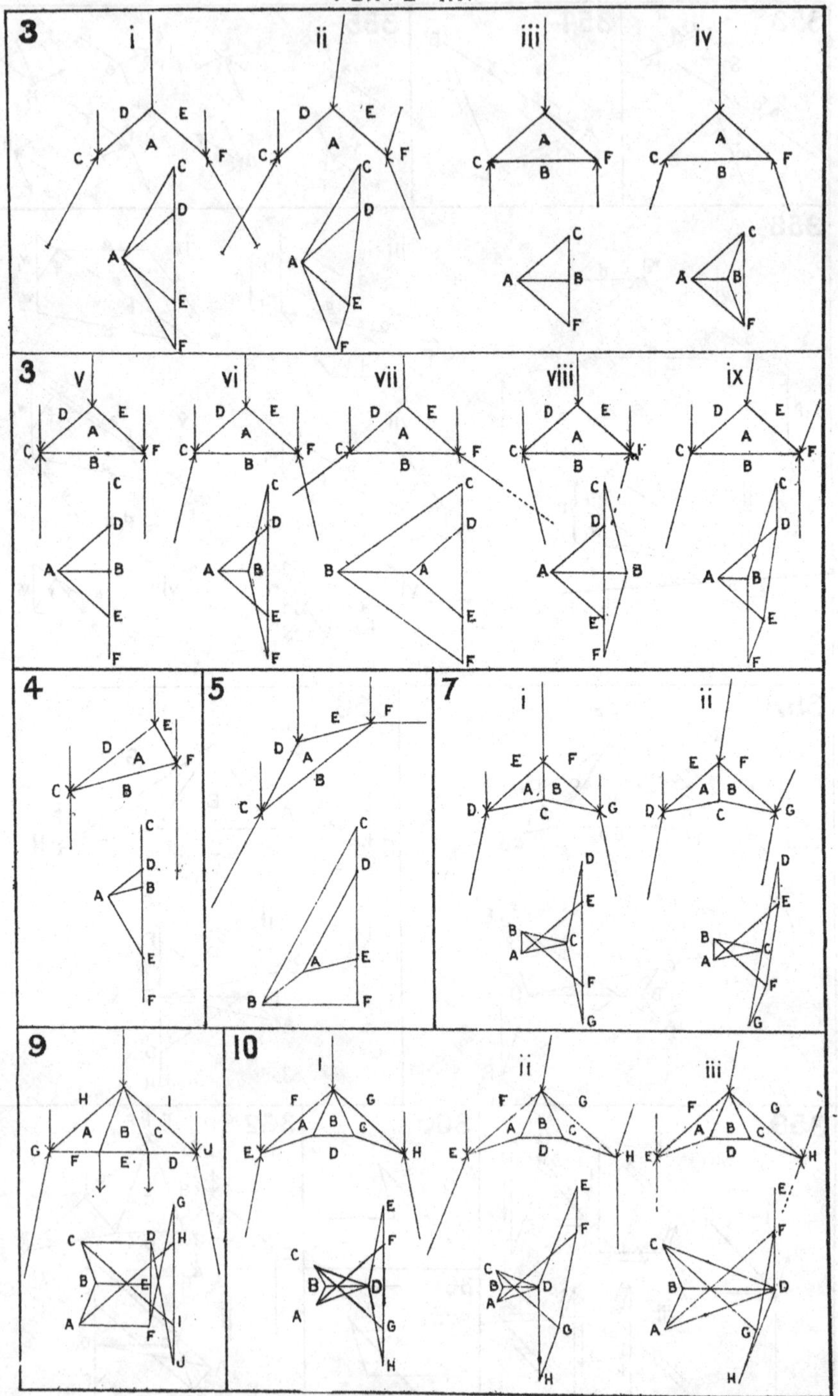

PLATE IV.

PLATE V.

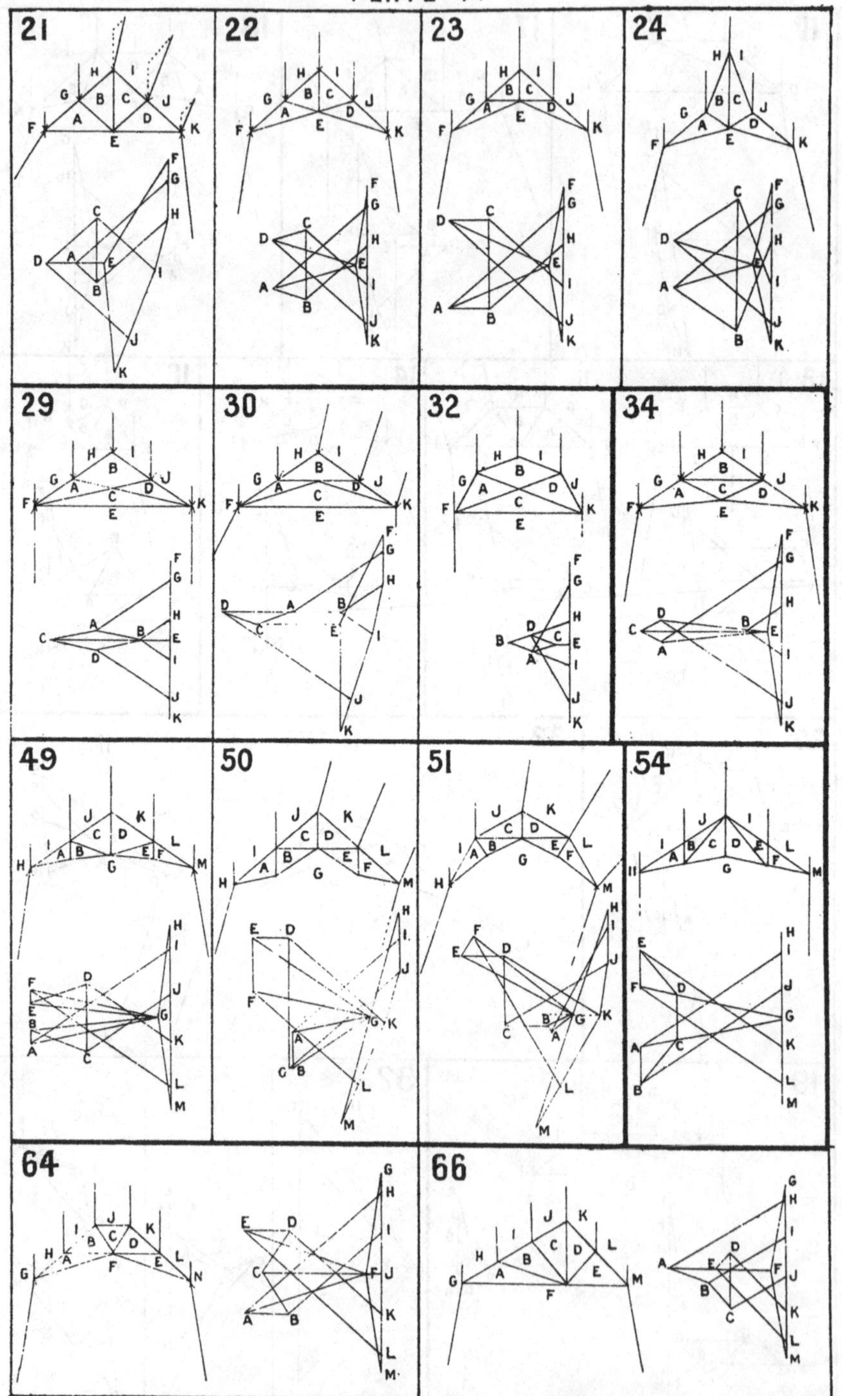

PLATE VI.

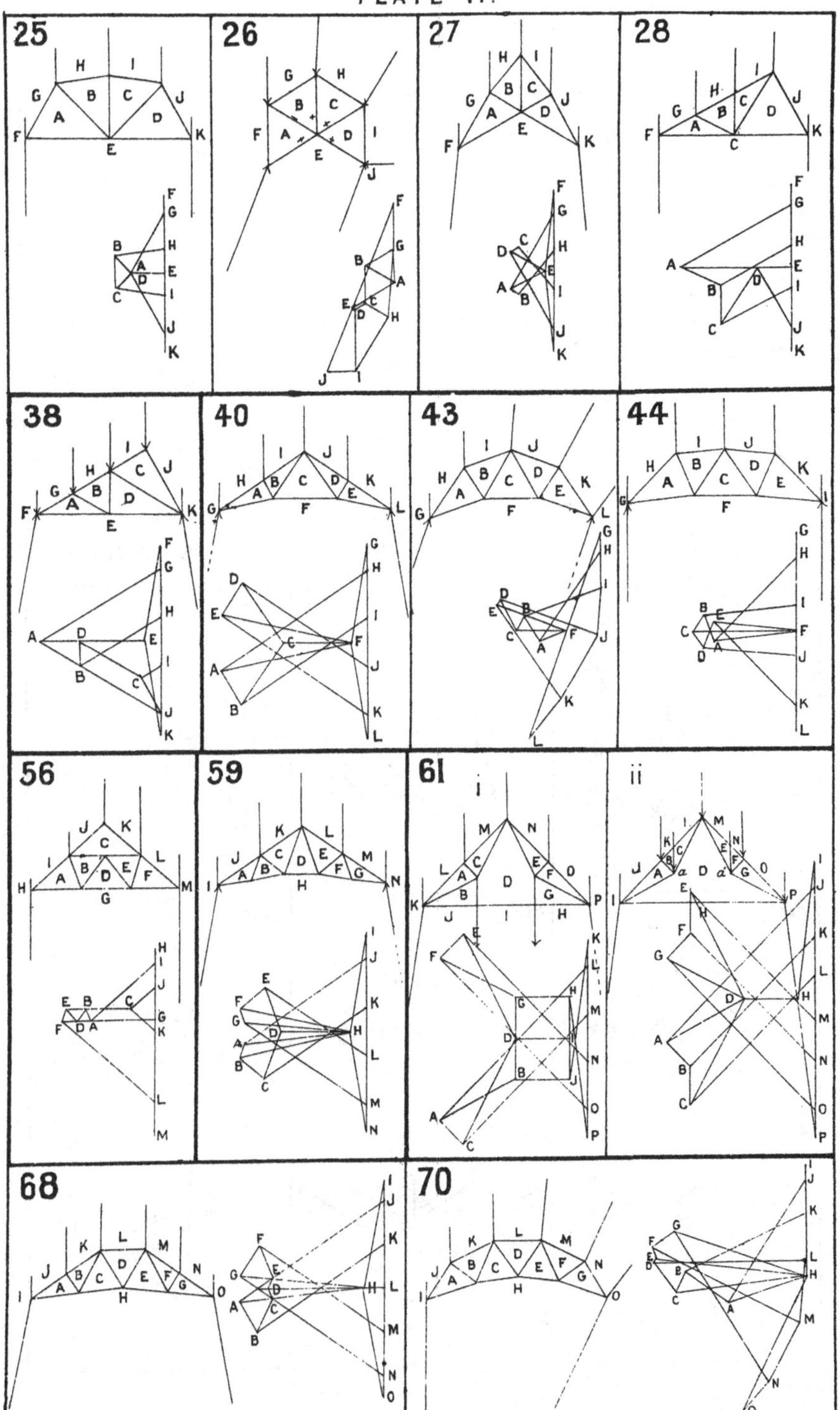

PLATE VII.

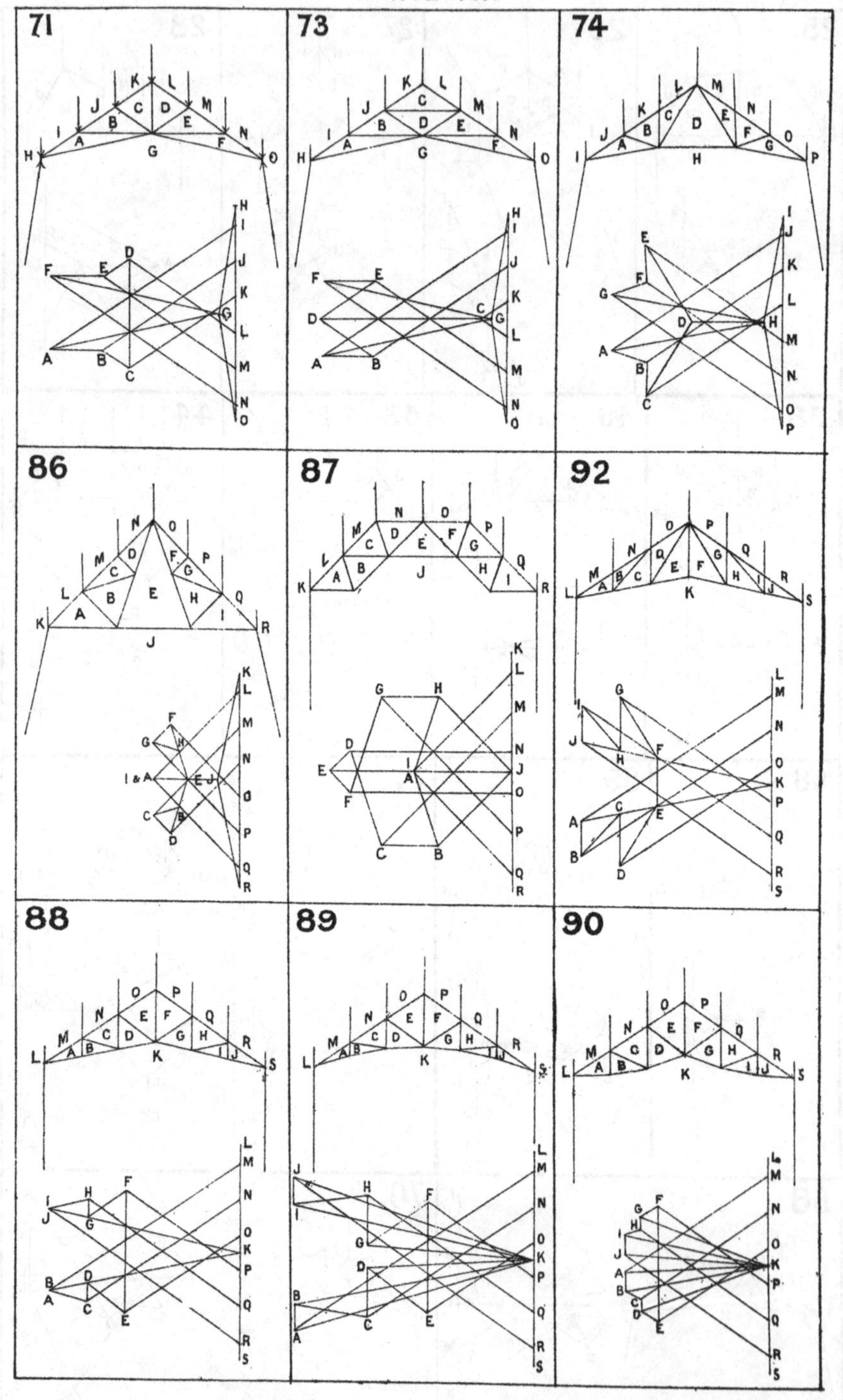

PLATE VIII.

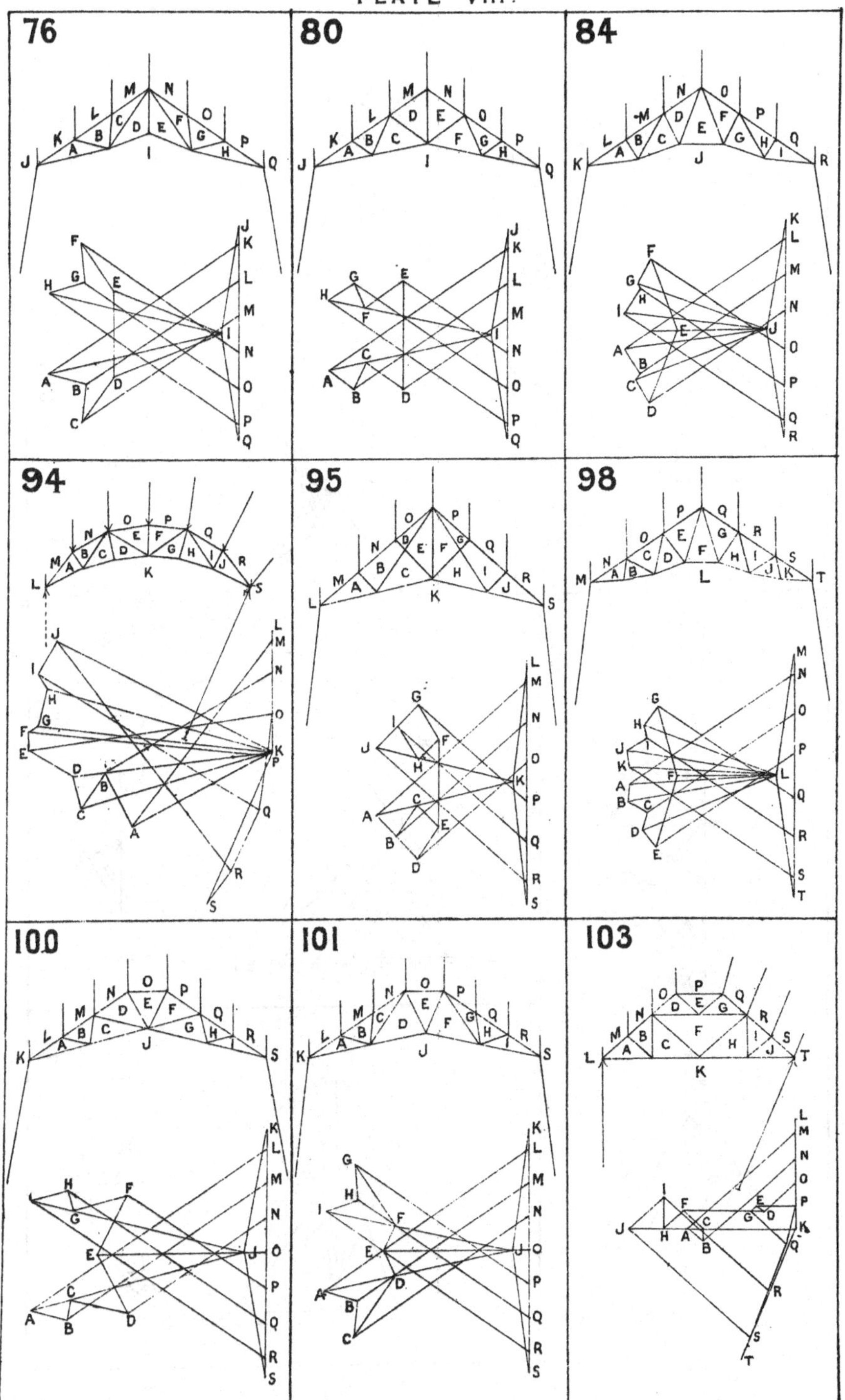

PLATE IX.

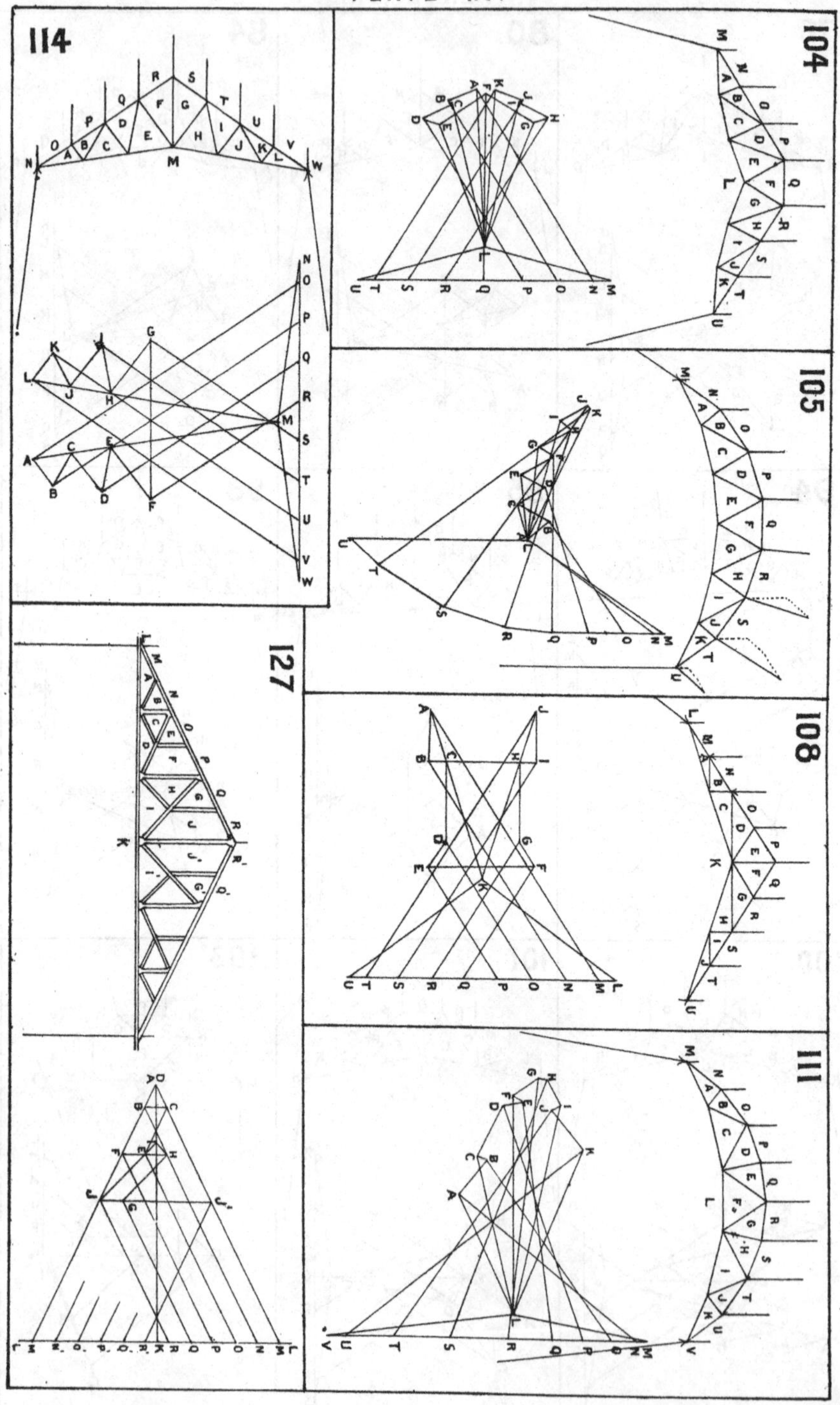

PLATE X

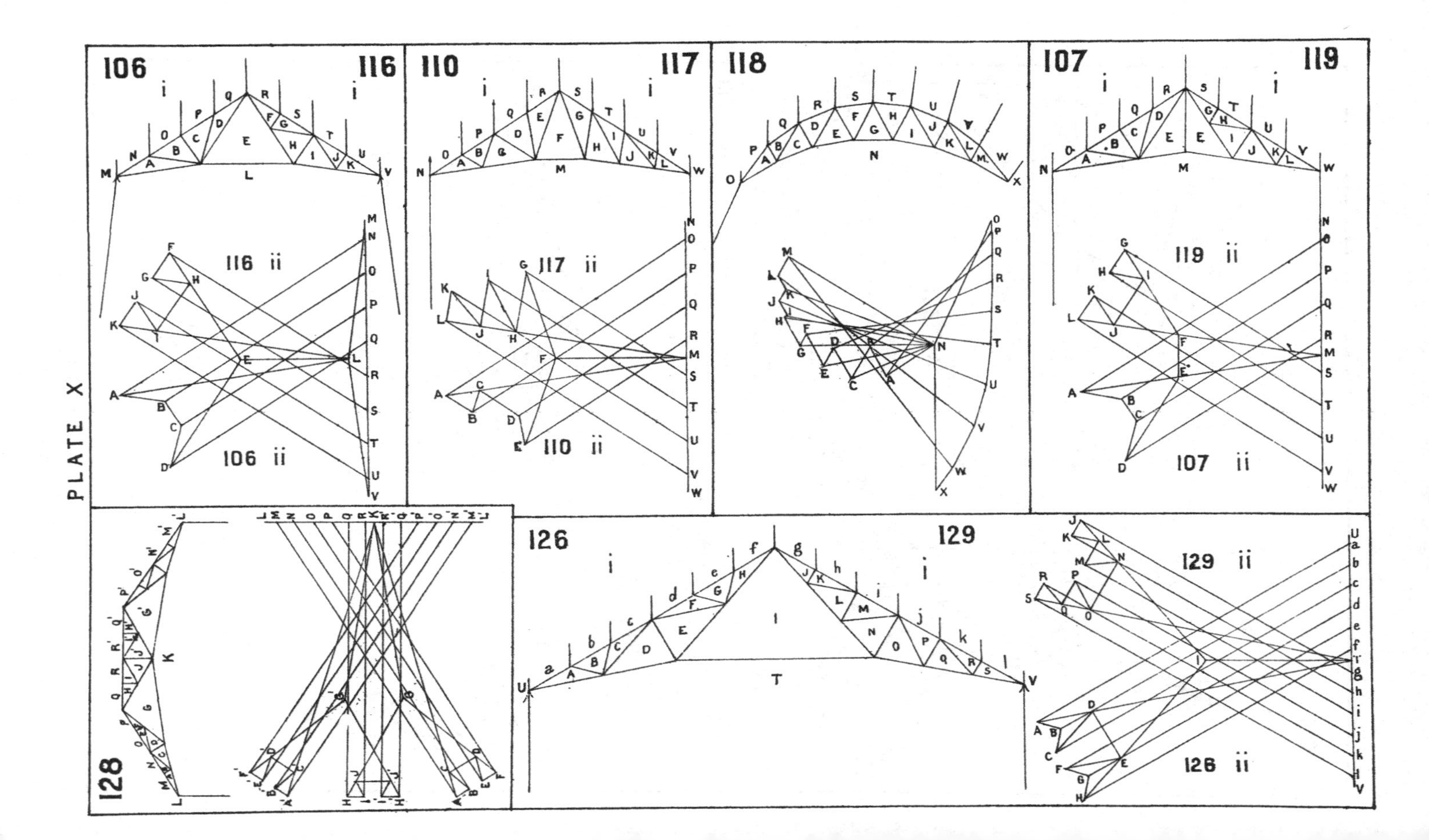
106 i
116 i
116 ii
106 ii
110 i
117 i
117 ii
110 ii
118
107 i
119 i
119 ii
107 ii
128
126 i
129 i
129 ii
128 ii

PLATE XI.

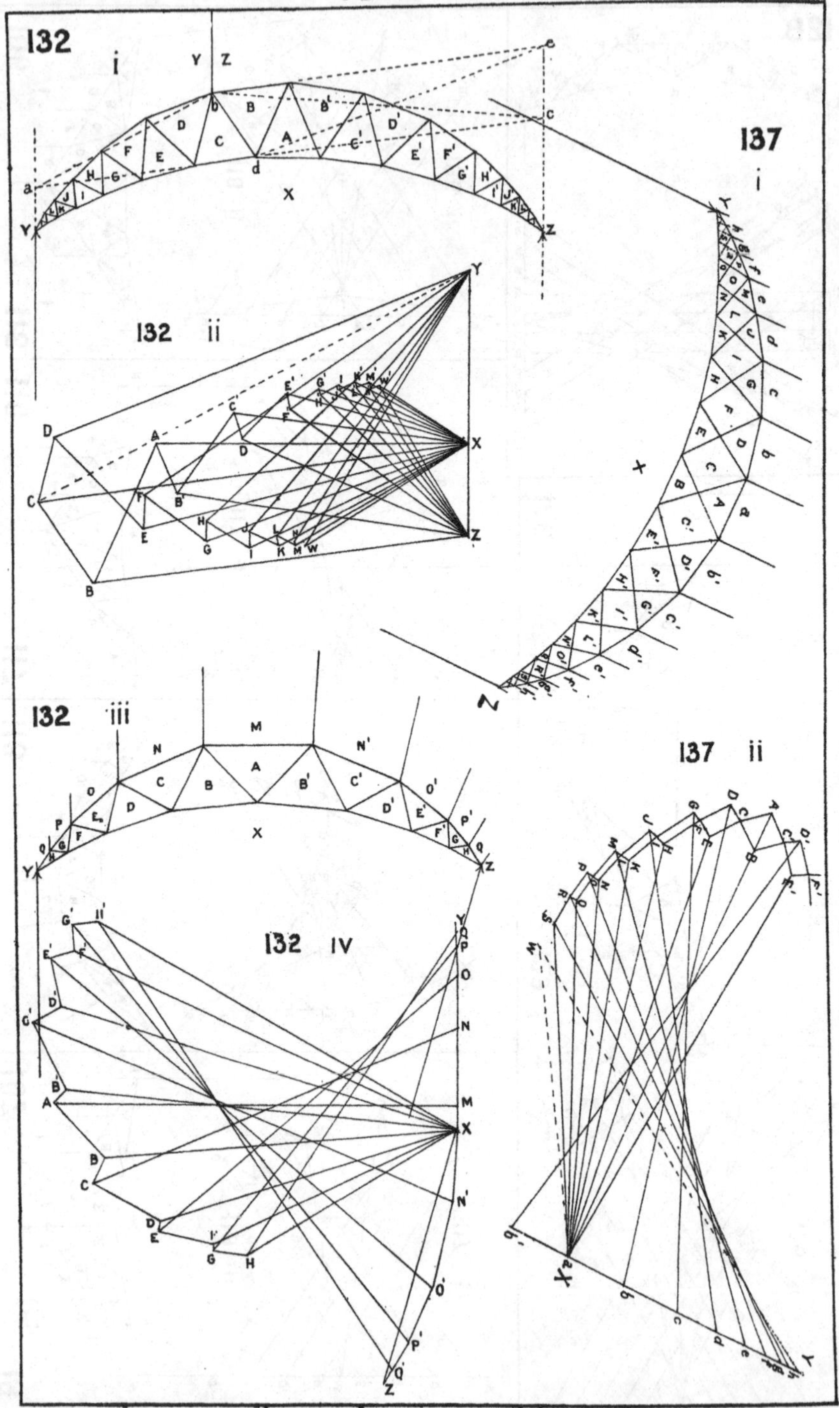

PLATE XII

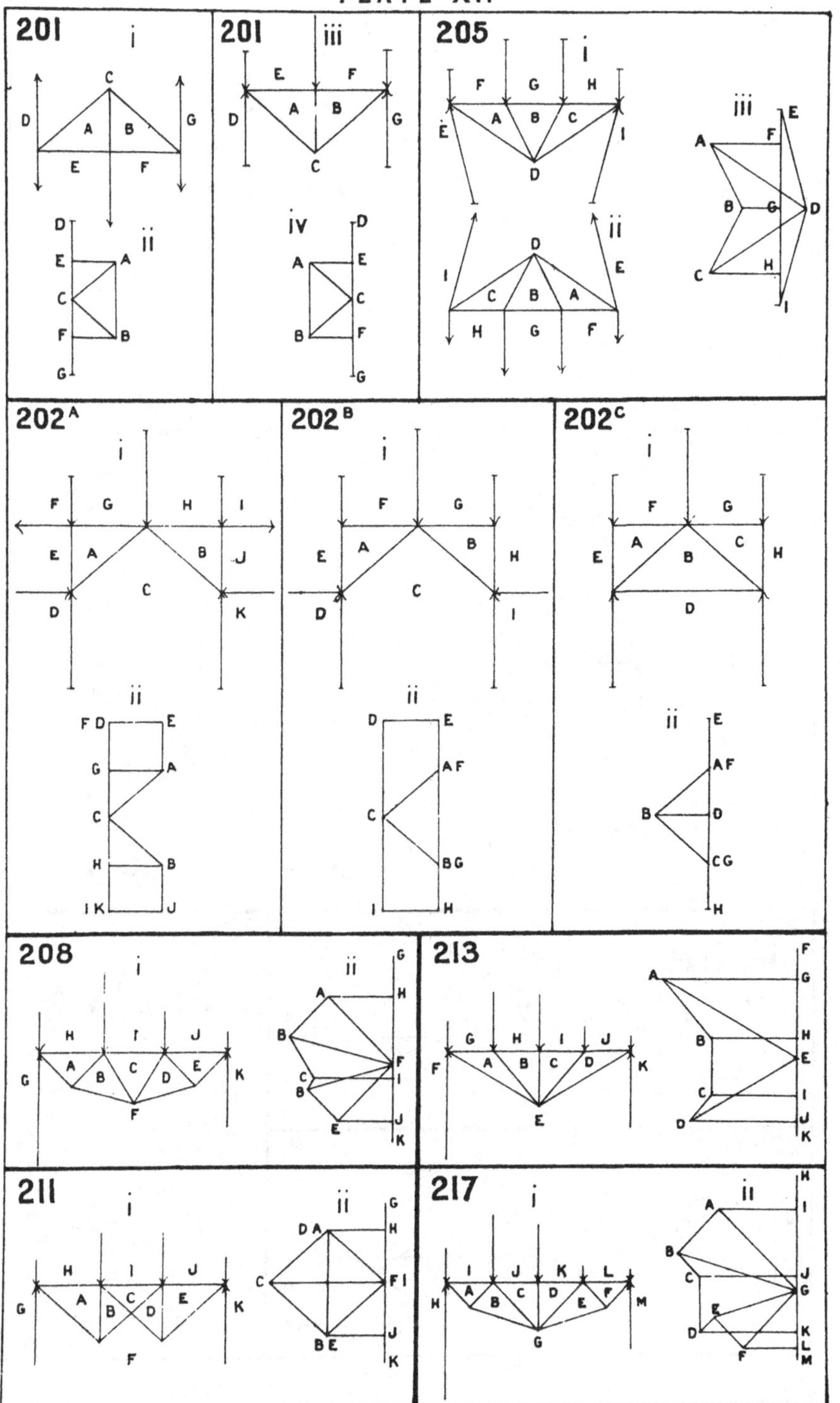

201
i
ii
201
iii
iv
205
i
ii
iii
202A
i
ii
202B
i
ii
202C
i
ii
208
i
ii
213
211
i
ii
217
i
ii

PLATE XIII

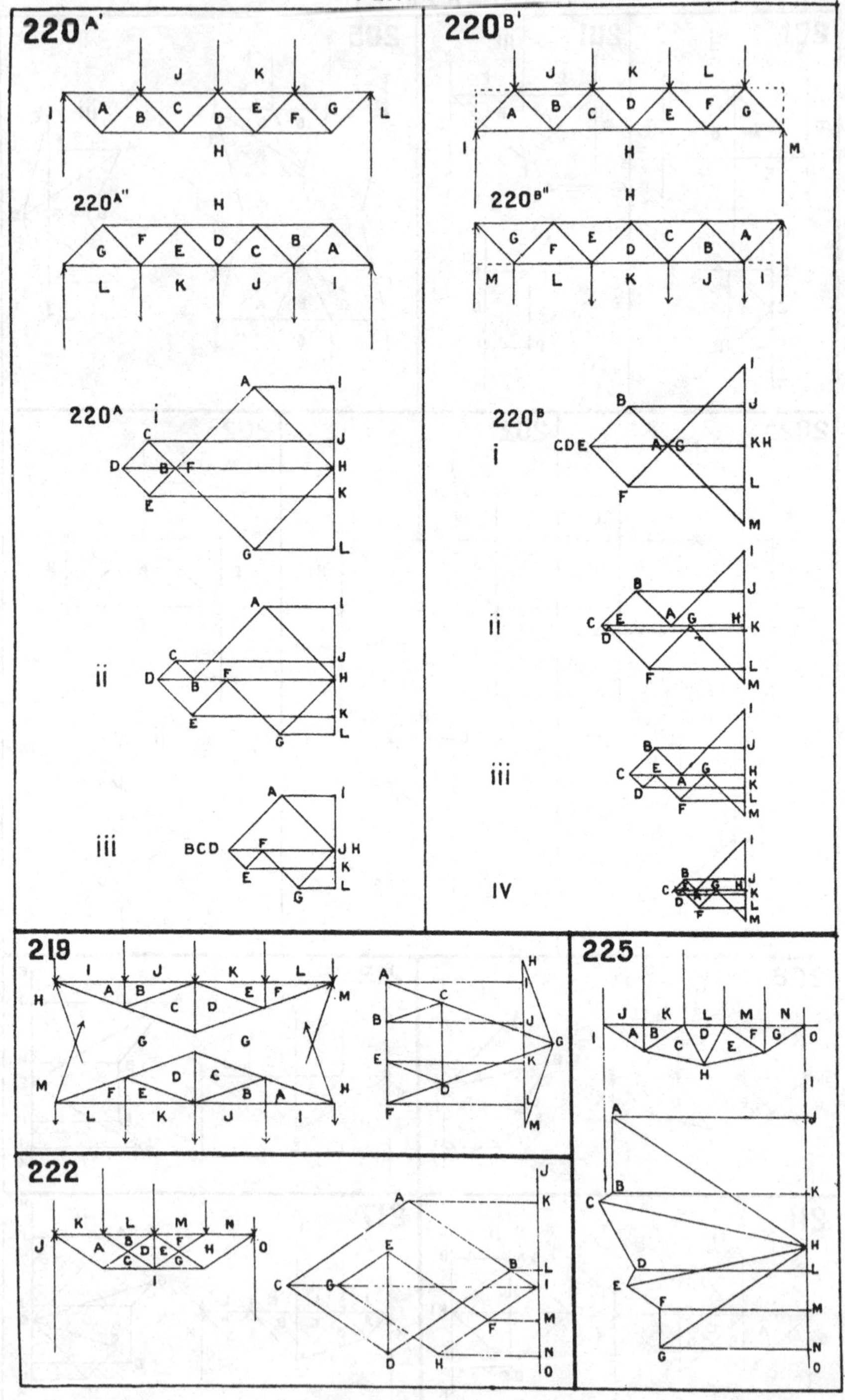

PLATE XIV.

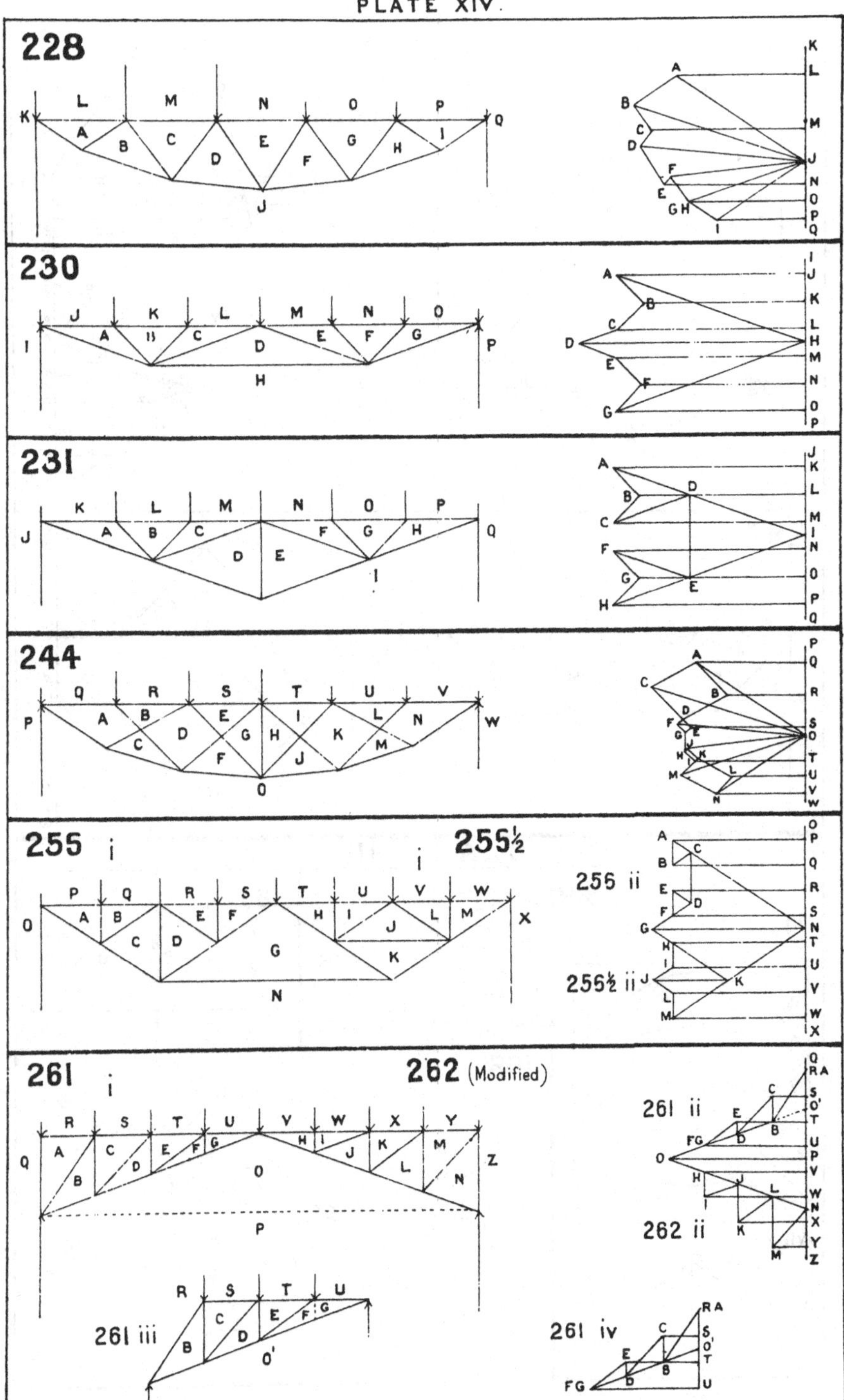
228
230
231
244
255
i
255½
i
256 ii
255½ ii
261
i
262 (Modified)
261 ii
262 ii
261 iii
261 iv

PLATE XV.

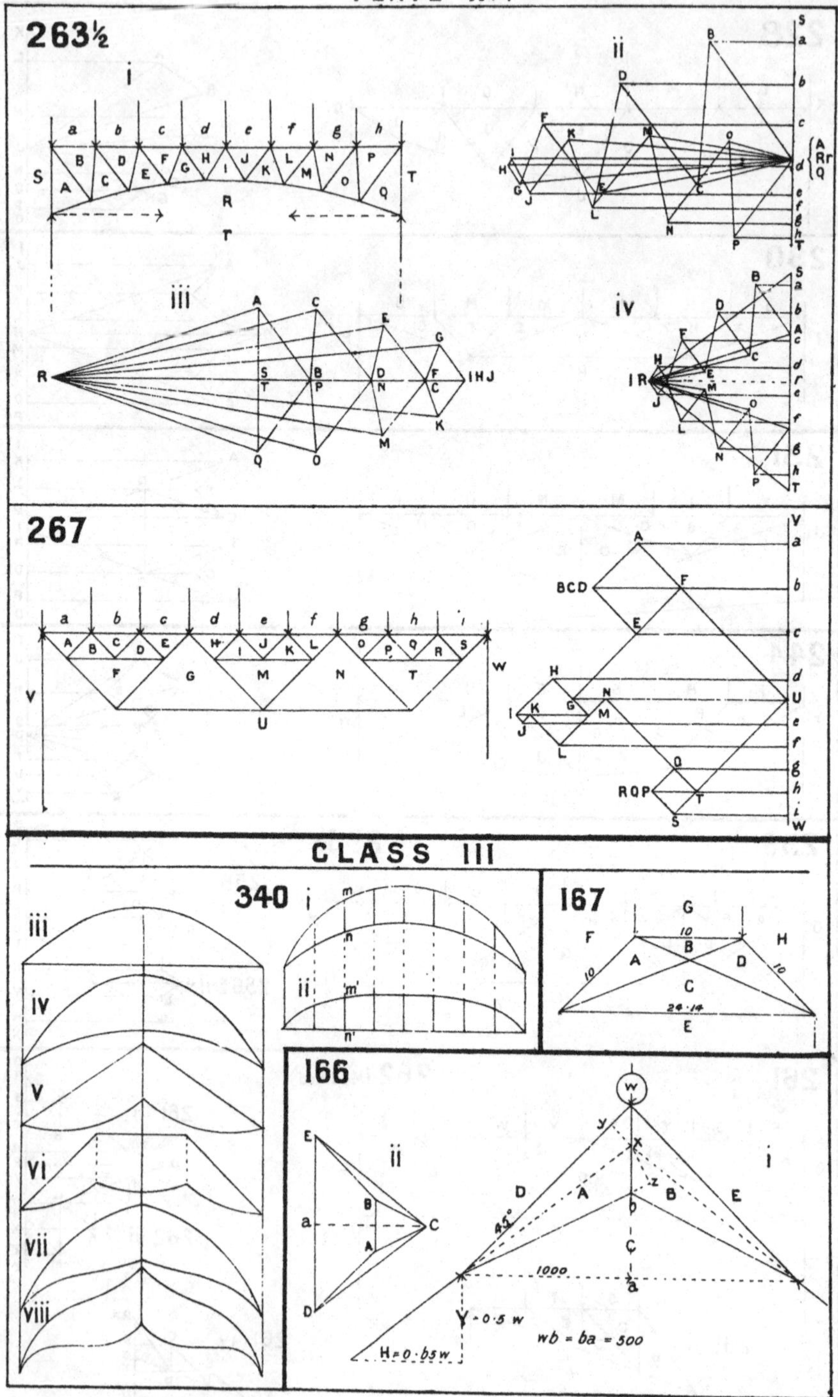

263½
i
ii
iii
iv
267
CLASS III
340
167
166
wb = ba = 500
H = 0·b5 w
Y - 0·5 w
1000
45°
24·14

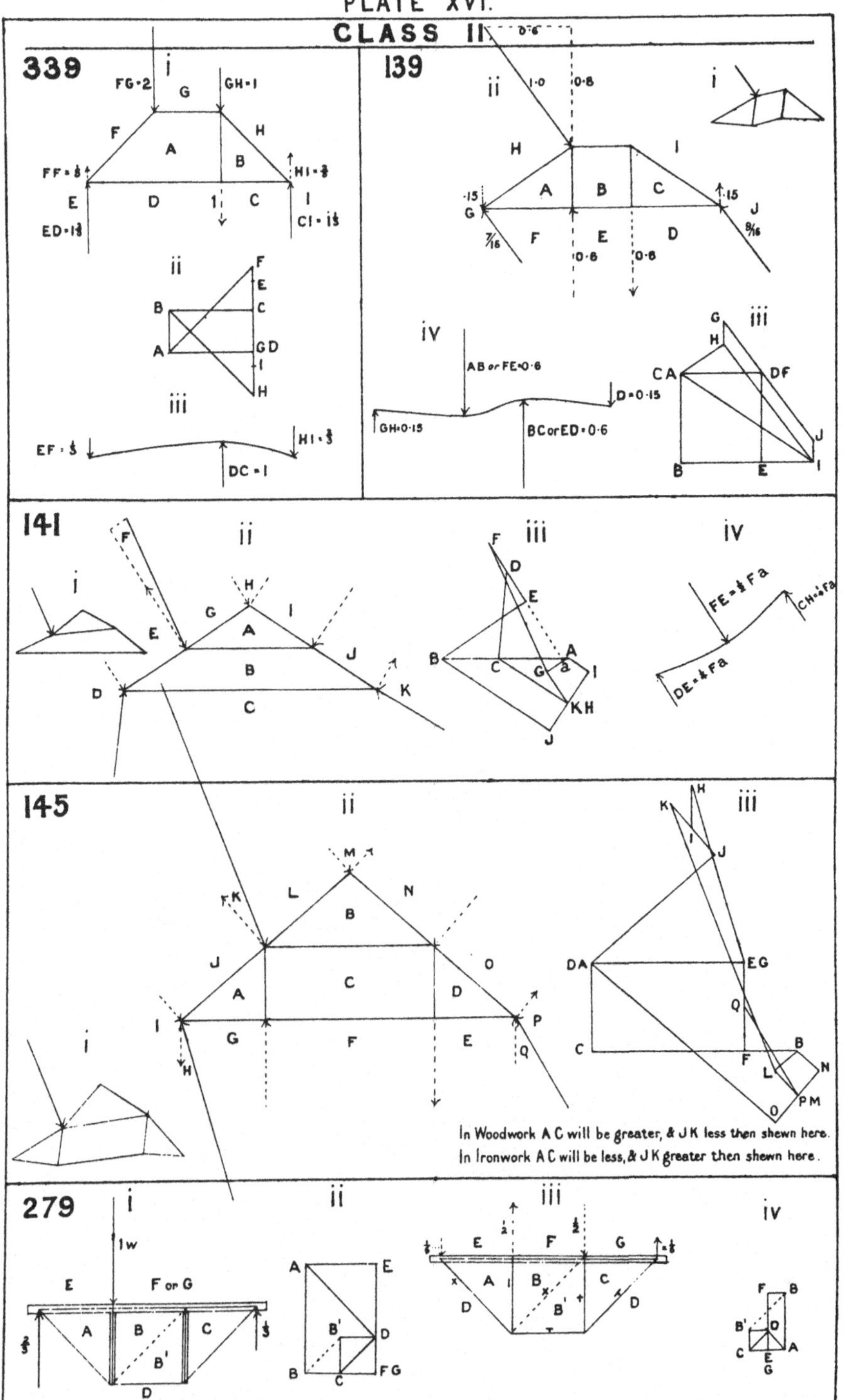
PLATE XVI.
CLASS II
339
i
FG=2
G
GH=1
F
A
H
B
E
D
C
ii
iii
DC=1
139
i
ii
iii
iv
AB or FE=0·6
D=0·15
GH=0·15
BC or ED=0·6
141
i
ii
iii
iv
145
i
ii
iii
In Woodwork A C will be greater, & J K less then shewn here.
In Ironwork A C will be less, & J K greater then shewn here.
279
i
1w
F or G
ii
iii
iv

Printed in the United States
By Bookmasters